M000036550

Iron and Manganese
Removal Handbook

Iron and Manganese Removal Handbook

Elmer O. Sommerfeld

Prepared for

Saskatchewan Environment and Resource Management

American Water Works Association
Dedicated to Safe Drinking Water

Copyright© 1999 American Water Works Association

No part of this publication may be reproduced or transmitted in any form or by any means, electronic or mechanical, including photocopy, recording, or any information or retrieval system, except in the form of brief excerpts or quotations for review purposes, without the written permission of the publisher.

Project manager and copy editor: David Talley
Production manager: Scott Nakauchi-Hawn
Cover design: Karen Staab

Library of Congress Cataloging-in-Publication Data:

Sommerfeld, Elmer O., 1937–
    Iron and manganese removal handbook / Elmer O. Sommerfeld
        p. cm.
    Includes bibliographical references and index.
    ISBN 1-58321-012-1
    1. Water--Purification--Iron Removal--Handbooks, manuals, etc. 2.
    Water--Purification-- Manganese Removal--Handbooks, manuals, etc. I. Title.

    TD466 .S62 1999
    628.1'666--dc21

                                                        99-047524

ISBN 1-58321-012-1
Printed in the United States of America

American Water Works Association
Dedicated to Safe Drinking Water

American Water Works Association
6666 West Quincy Avenue
Denver, CO 80235
(303)794-7711

Table of Contents

Figures

Tables

Preface

This book is an updated version of the Iron and Manganese Removal Workshop Manual originally created in 1984 by the Saskatchewan Environment and Resource Management Department for water treatment plant operators. In the intervening years, major progress has been made in understanding processes for removing iron and manganese from both groundwater and surface water sources. Also, however, much remains to be learned about (1) how other raw-water constituents interfere with typical treatment methods, (2) how those interferences can be readily and easily identified, and (3) how they can be reduced or eliminated in a cost-effective manner (often a problem in very small communities). Other relevant issues include filter and process auditing and pilot testing, which have made major contributions toward rehabilitating existing iron and manganese removal plants, updating removal processes, and inspiring new plant designs.

Although it was developed as a training tool, this handbook will be a useful reference for anyone interested in removing iron and manganese from potable water, as well as the basic operation of water treatment and filtration plants. The handbook was compiled under the direction of the following individuals from the Municipal Branch, Saskatchewan Environment and Resource Management:

Blaine W. Ganong, Manager, Municipal Assessment and Certification Section

Gus Feitzelmayer, Operator Training and Certification

The personal workshop files of Peter Thiele, Supervisor of Municipal Environment Officers, Municipal Branch, Saskatchewan Environment and Resource Management, offered valuable help in compiling the information in this handbook. The following individuals contributed by writing certain chapters or sections, by suggesting appropriate graphics, and by offering reviews, critiques, and other professional assistance:

J.B. Hambley, B.App.Sc., Chemical Engineering, P.Eng.
Founder, Anthratech Western Inc.
Executive Officer, Hambley Engineering Inc.

Mike F. Adkins, B.Sc. (Eng) Civil Engineering, P.Eng.
Manager, Environmental Engineering Division
Reid Crowther & Partners Limited

David M. Hambley, A.Sc.T.
CEO and Operations Manager
Anthratech Western Inc.

G.A. Irvine, Ing. Qco., M.Sc., P.Eng.
Senior Project Engineer, Environmental Engineering Division
Reid Crowther & Partners Limited

William R. Knocke, Ph.D., P.E.
Head, Department of Civil Engineering
Virginia Polytechnic Institute and State University

David O. Hyndman, B.A.
Teacher: English, History, Computer Education, Information Processing
Spiritwood High School

K.M. Delsnider, B.Sc., P.Eng.
Manager, Civil Engineering Division
Reid Crowther & Partners Limited

Elmer O. Sommerfeld
Water Process Troubleshooter
Anthratech Western Inc.

The authors have taken measures to keep the content of this handbook readily understandable by operators of small water treatment systems. Although highly scientific or technical subject matter has been avoided, all basic concepts of iron and manganese removal have been included. Chapters and sections on mathematics, chemistry, physics, microbiology, hydraulics, and hydrology have been included, as any operator should have a basic understanding of these subjects to maintain efficient operation of a water treatment plant.

In practical activities, an operator may need to use metric, Imperial, and US customary units as interchangeable measures. Operating manuals for new package filtration plants manufactured in the United States use US customary units or varying combinations of US and metric units. Technical data sheets issued currently by North America's only manufacturer of manganese greensand (located in New Jersey), show volumes in US gallons, temperatures in degrees Fahrenheit, and pressures in pounds per square inch (psi).

This handbook emphasizes metric units. It shows the currently accepted standard for each area of the industry will be shown in brackets. A table of common conversions is included at the end of the book.

Elmer O. Sommerfeld
Chief Writer

Chapter 1

Introduction

The Hydrologic Cycle

Hydrology is the science concerned with the distribution of all the earth's waters and their continual movement from oceans and lakes to the atmosphere, from the atmosphere to the land as surface water, and from the land back to the oceans and lakes (see Figure 1-1). Over 97 percent of all the earth's water is contained in the oceans. The remaining 3 percent is distributed among the atmosphere as water vapor; on land as fresh water, ice, and snow; and as groundwater beneath the earth's surface.

Some 75 percent of all fresh water is held by the polar ice caps and glaciers. Most of the fresh water balance lies in aquifers (water-bearing layers of rock, sand, and gravel). A scant one-third of 1 percent of all fresh water is in lakes and rivers, about 20 percent of this world total in the Great Lakes between Canada and the United

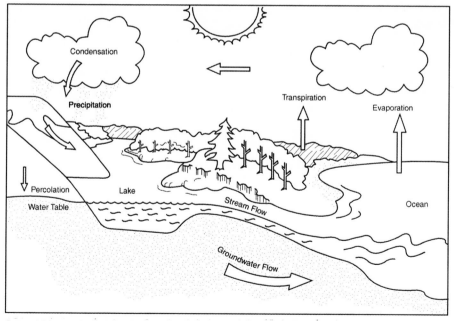

Source: A Primer on Fresh Water, Canada, 1994. Reproduced with the permission of Environment Canada, 1999.

Figure 1-1 The hydrologic cycle

States. Not all water moves through the hydrologic cycle in the same amount of time. Some is evaporated from the oceans, falls as rain on coastal areas, and returns to the oceans within days. Other precipitation takes years to complete the cycle, while still other precipitation never reaches the oceans and lakes, completing its cycle by returning to the atmosphere through evaporation and/or transpiration (movement of water from the soil to the atmosphere through the actions of growing plants).

About 25 percent of the earth's fresh water is stored beneath the planet's surface, where it can spend thousands of years. Of this total, only a small amount is located in easily accessible aquifers that can produce worthwhile volumes of water on an ongoing basis. The important consideration in access to this water, then, is not how much can be drawn from a source, but how quickly the hydrologic cycle can replenish it. Without replenishment, supplies would soon be depleted. Therefore, reservoirs are often constructed to hold adequate supplies for dry seasons, low river flows, and winter months. Good wells, on the other hand, can usually produce reliable, consistent flows day in and day out for years.

Figure 1-2 shows a hypothetical groundwater system. Although water sometimes appears to be at rest, it rarely is. As the figure shows, water is always moving within an aquifer, recharging an aquifer, flowing into a lake, or being contaminated by synthetic products or saltwater. Heavy rainfall, rivers flooding their banks, or drought can all affect raw-water quality in the short term. Often, seemingly small quality changes in raw water significantly affect a treatment or iron (Fe) and manganese (Mn) removal process.

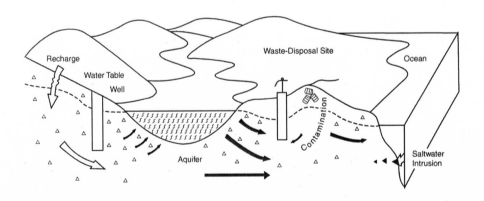

Source: A Primer on Fresh Water, *Canada, 1994. Reproduced with the permission of Environment Canada, 1999.*

Figure 1-2 Groundwater System

Iron and Manganese: Nature and Occurrence

These two elements, both relatively abundant in the earth's crust, can cause serious aesthetic problems in potable water systems. If iron were always in the simple form of Fe and manganese were always simple Mn, removal would be a simple process, but such is not the case.

Simple soluble reduced iron is ferrous iron (Fe^{2+}), and simple soluble reduced manganese is manganous manganese (Mn^{2+}). When oxidized, the iron changes to ferric iron (Fe^{3+}), and manganese often (but not always) changes to the insoluble Mn^{3+} or Mn^{4+} forms. These concepts are explained in detail in chapter 2. Both Fe and Mn can form combinations with many other elements and compounds. These other combinations generally do not prevent removal, however, and often they do not inhibit the oxidation/flocculation/sedimentation/filtration process.

Fe and Mn usually occur only in wells and impounded surface supplies (reservoirs). Flowing waters (rivers and streams), even if they normally do not contain significant levels of Fe and Mn, are certain to contain higher levels of these elements and related compounds after being impounded. The amount that goes into solution depends on the character of the soil and the amount of plant life. Decomposition of organic matter (algae, leaves, and other plant material) in the lower sections of a reservoir may result in anaerobic conditions (i.e., near zero oxygen) under which Fe and Mn compounds in those zones are converted to soluble compounds. Groundwaters with intolerable amounts of Fe and Mn contain little or no dissolved oxygen and are often relatively high in carbon dioxide. The absence of dissolved oxygen indicates anaerobic conditions, and high carbon dioxide content suggests extensive bacterial oxidation of organic matter. The presence of hydrogen sulfide is another indicator of the same condition.

Attempts to remove Fe and Mn from impounded surface waters encounter a further challenge at least twice a year in much of North America. Cold water at the top of a reservoir sinks to the bottom, while warmer water at the bottom rises to the top following spring breakup and again just before freeze-up; this sequence is called a *water inversion* or *turnover*. The water rising to the top brings with it soluble compounds of Fe and Mn. As wind and wave action mixes oxygen from the air with the water, Fe and Mn are gradually oxidized (an increase in positive valence resulting from the presence of oxygen, as the glossary explains). The oxidized elements precipitate, or separate out of solution. They sink to the bottom again as solids and are brought back into solution where oxygen is absent and carbon dioxide has been formed. While some of the Mn^{2+} is very slowly oxidized in this way, action of certain bacteria causes a much faster rate of oxidation. The scientific name for this process is *microbially mediated oxidation*. Microbially mediated Mn^{2+} oxidation can make a very important contribution to manganese cycling in a reservoir.

An operator faces the challenge of learning how to read the signs of an inversion and then to adjust the treatment process (usually through a combination of actions, such as adding chemicals, shortening filter runs by backwashing more often, etc.) to handle the increased levels of Fe and Mn in the raw water coming into the treatment plant. If time and technology permit, measurements of dissolved oxygen should be taken, and a temperature profile should be developed to assist in identifying stratification. Operating experience helps to position the operator to know just when to prepare for increased levels of Fe and Mn in the raw water. Because these occurrences do not happen at the same time every year, a strong case can be made for historical record keeping. Guided by records showing dissolved oxygen levels, stratification data, weather data, time lapses between events, and dates of events, the operator can respond promptly when the occurrence repeats itself. Gases are usually released as a part of the total water inversion occurrence; however, by the time odors can be detected, Fe and Mn have long ago been released.

During much of the year in temperate climates, water nearest the surface of a reservoir is least likely to contain Fe and Mn. Therefore, the inlet to the water treatment plant feed line should ideally move up and down, so that water can always be taken from a level below surface plant growth, such as algae, but above the zone lacking oxygen.

Amounts of Fe and Mn vary widely across North America, and they can vary widely within a small geographical area, depending on aquifer structure, well depth, surface soils, rock formations, plant types and rates of growth, and many other factors. Some reservoirs accumulate significant amounts of Fe and Mn in short periods of time, while others build up concentrations over many years.

Effects of Iron and Manganese in Potable Water Systems

The problems resulting from Fe and Mn in potable water addressed in this section are not related to health. Fe and Mn in solution do not commonly rise to amounts considered hazardous to physical well-being in raw-water sources destined for potable use, although some researchers claim that many metals (including manganese) have a negative impact on the body over many years of ingestion.

Instead, these metals cause aesthetic problems. Yellowish to reddish-brown stains result from water containing ferric iron oxides–hydroxides. Manganese-bearing waters with no iron produce dark brown or black stains. Combining the two metals results in stains that range from light brown to black. Initial complaints usually cite stained laundry; later complaints describe brown or dirty water; finally, additional buildup may allow chunks sloughed off in the distribution lines and broken up by the movement of the water to pour from customers' taps.

A treatment and filtration process can control customer complaints of laundry spotting and/or dirty water by producing water with less than 0.3 mg/L iron and less than 0.05 mg/L manganese. Reduction of metals to these levels is not a guarantee the operator will receive no brown-water complaints. Reduction of iron to 0.1 mg/L and manganese to 0.02 mg/L is a positive step, as is maintenance of the distribution system in a clean condition. Sporadic complaints or repeated complaints from the same consumer usually result from water heaters that need cleaning, softener beds heavily coated with iron and manganese oxides, and/or overuse of cleaning and laundry products high in chlorine.

Presence of Fe and Mn promote the growth of crenoforms, which are stringy-looking, microscopic organisms. Crenoforms (with scientific names including *Crenothrix, Leptothrix, Sphaerotilus,* and *Gallionella*) congregate in piping to form heavy, jelly-like, stringy masses that can impair the hydraulic (water) carrying capacity of an entire system. Allowing the formation of these organisms through inadequate removal of Fe and Mn and inadequate disinfection of filtered water at the water treatment plant is almost sure to result in a substantial cost in time and money to flush and/or swab distribution lines.

According to Cullimore (1993), "It was found during that research phase that 95 percent of the groundwaters tested in Saskatchewan, Canada, were positive for IRB (iron-reducing bacteria). Microscopic examination found that, of the sheathed and stalked IRB, *Crenothrix, Leptothrix, Sphaerotilus,* and *Gallionella* were frequently dominant types. Given the universal nature of the presence of IRB it becomes more critical to appreciate their relative aggressivity." For more information on the effects of these and other living creatures on iron and manganese removal, see chapter 4 on microbiology.

Since 1962, the accepted industry standard for maximum allowable limits of Fe and Mn in treated potable water has been 0.3 mg/L (milligrams per litre) iron and 0.05 mg/L manganese. A good general rule, but not a universal one, is to focus on manganese removal, because iron usually falls below 0.3 mg/L when manganese has been reduced below 0.05 mg/L. This tendency was confirmed by a study conducted in 1993 of Fe and Mn removal plants in the province of Alberta, Canada.

Chapter 2
Chemistry of Iron and Manganese Removal

This chapter provides an overview of issues related to iron (Fe) and manganese (Mn). The removal of Fe and Mn below targeted levels (i.e., 0.3 mg/L Fe and 0.05 mg/L Mn) can be a difficult objective, because so many factors affect the treatment and removal process. These issues range from other raw-water constituents to flow rates, pH, water temperature, etc. Topics introduced in this section are covered in greater detail in other chapters. If the information presented here is insufficient for your needs, search the index for specific references.

Water Chemistry

Before iron and manganese can be removed from water by any method, some chemical change must take place. Chemical changes are also usually responsible for any difficulty encountered in removing Fe and Mn.

A treatment plant operator need not memorize the periodic table of the elements, understand atomic structure, or know how to calculate molecular weight, but it is vitally important for an operator to understand some basic changes that take place when certain chemicals come into contact with certain other elements and compounds. For example, the presence of organic carbons, hydrogen sulfide, or ammonia often indicate a potential for interference with Fe and Mn removal processes, thus an operator should have a basic knowledge of organic carbons and ion exchange principles.

A *chemical element* may be defined as a chemical substance that cannot be broken down into simpler substances by ordinary chemical change. Water, a chemical *compound*, easily breaks down further into hydrogen and oxygen (which are chemical elements). Hydrogen and oxygen cannot break down into still simpler substances without altering their basic identities.

The smallest part of an element is an *atom*; it has all the chemical properties of the element in which it exists. Smashing an atom produces new elements of lighter weight and releases a tremendous amount of energy. Atom smashing (*fission*) is the basis for atomic energy technology.

Chemical Symbols, Common Formulae

In the language of chemistry, each element is represented by a symbol. A combination of symbols to characterize a chemical compound is called the *formula* for that compound. A formula gives all the elements in the chemical compound and the number of atoms of each:

- Symbol for hydrogen: H
- Symbol for oxygen: O
- Formula for water: H_2O; the compound water has two atoms of hydrogen combined with 1 atom of oxygen. Note that single atoms are not numbered.
- Symbol for sodium: Na
- Symbol for chlorine: Cl
- Formula for sodium chloride (ordinary table salt): NaCl; the compound has only one atom of each element.
- Symbol for iron: Fe
- Formula for ferric oxide (red rust): Fe_2O_3; the compound has two atoms of iron combined with three atoms of oxygen.

Ionization

Many elements combine with each other on contact. This change can happen slowly, rapidly, or explosively, often under the influence of heat, moisture, or catalytic agents (substances that promote reactions but do not themselves undergo any change during the reactions). Many chemical reactions take place in the presence of water, because water can split some molecules into positively or negatively charged atoms called *ions*. Naturally, this process is known as *ionization*. A molecule that produces ions when dissolved in water is said to *ionize*.

Example: $NaCl \rightarrow Na^+ + Cl^-$. The electrically charged atoms are ions.

Positively charged ions are called *cations* (pronounced "kat'i'ens"), and negatively charged ions are *anions*.* The charges must be equal in strength and opposite in sign, otherwise the solution itself would have a charge, which is impossible.

When a molecule ionizes in water, the charges must equalize even when the molecule combines more atoms of one kind than another. For example, $FeCl_3$ (ferric chloride) ionizes like this: $FeCl_3 \rightarrow Fe^{+3} + 3Cl^-$.

Some molecules ionize to yield groups of atoms, consisting of positively charged cations and negatively charged anions. For example, Na_2SO_4 (sodium sulfate) ionizes like this: $Na_2SO_4 \rightarrow 2Na^+ + (SO_4)^{-2}$, leaving two sodium cations and one sulfate anion. Combinations of atoms in a unit, such as $(SO_4)^{-2}$ are called *radicals*.

---

* These designations are in common use today when referring to the electrical charges carried by polymers, which are used as coagulants and filter aides. To increase their efficiency, many polymers today are either cationic or anionic, and users can specify the strength of the electrical charge each carries.

Ion Exchange

A water softener is a common household appliance that removes water hardness, or softens the water, by ion exchange. Hardness is substantially composed of calcium (Ca) and magnesium (Mg). (Note the difference between magnesium and manganese.)

Soap is typically a water-soluble sodium salt of a long-chain organic acid. If the sodium in the soap is exchanged for calcium or magnesium, an insoluble calcium or magnesium soap is formed. Thus, waters high in calcium and/or magnesium are hard to wash with due to the formation of insoluble Ca/Mg soap. That is why the term *hardness* arose.

So, when calcium and magnesium ions are exchanged for sodium ions, the water is said to be *softened*. Since each ion has a positive charge, the process is not ionization (as previously described), but ion exchange; a more scientific term would be *cation exchange*.

Removal of hardness lasts only as long as sodium (Na)* is left on the exchange medium (typically, synthetic zeolite resin beads). Depletion of the sodium requires regeneration of the zeolite bed by passing a strong solution of either salt (NaCl) or potassium chloride (KCl) through it to replace the calcium and magnesium exchanged.

Zeolite softening of potable water has its limitations. The calcium and/or magnesium removed is replaced with sodium. High-sodium water is bad for people with hypertension, however, and poor for plants and lawn watering.

Iron and manganese are removed in a manner similar to this process for calcium and magnesium. Success requires caution to prevent oxidation of iron and manganese ahead of a cation exchange softener, or the precipitates of iron and manganese could foul (or coat) the resin beads.

Ion exchange can also remove all other cations, producing water of almost any quality. By regenerating with hydrogen compounds instead of sodium compounds, cation-free water can be produced. Further, the other ions in the water, anions such as sulfate and chloride, can be replaced by the hydroxyl ion (OH^-). In this way, deionized water can be produced, which is ionically the same as distilled water.

pH Value

Water ionizes to a slight degree, producing both hydrogen and hydroxyl ions, as follows:

$$H_2O \rightleftarrows H^+ + (OH)^-$$

Water may be described as both an acid and a base (alkaline), because it produces

---

* In many areas, sodium has been replaced by potassium, so the softened water can be consumed without adverse health effects and without negative effects on natural water and soil environments.

both hydrogen and hydroxyl ions. Since these ions are present in identical concentrations, however, pure water is neutral—it has no charge. But, when water ionizes, the extent of the ionization depends on the relative concentration of hydrogen and hydroxyl ions. A complicated mathematical formula for this phenomenon produces Table 2-1, which describes the relative acidity or alkalinity of any water sample. The pH scale goes from 0 to 14, neutral water having a pH of 7.

Careful examination of Table 2-1 discloses several important facts:

1. The highest hydrogen ion concentration corresponds to the lowest pH.
2. The lowest hydrogen ion concentration corresponds to the highest pH.
3. Neutrality at pH 7 defines the midpoint in the scale; pH values lower than 7 represent hydrogen ion concentration above neutrality (acidity), and pH values higher than 7 represent hydrogen ion concentration below neutrality (alkalinity).
4. In practice, pH values can fall between whole numbers. For example, one water's pH could be 6.1, another water's pH 6.2, and still another water's pH 9.5.

This variation shows once again the importance of ionization, the major role played in chemistry by hydrogen and hydroxyl ions, and the need for an operator to know just how acidic or basic (alkaline) a water is.

Table 2-1 Ionization and pH value

A Grams of H+ ions per litre	B Reciprocal of A	Log B	C (pH)
1.0	1.0	0	
0.1	10	1	
0.01	100	2	
0.001	1,000	3	0–6: Acid
0.0001	10,000	4	
0.00001	100,000	5	
0.000001	1,000,000	6	
0.0000001	10,000,000	7	Neutrality
0.00000001	100,000,000	8	
0.000000001	1,000,000,000	9	
0.0000000001	10,000,000,000	10	
0.00000000001	100,000,000,000	11	8–14: Alkaline
0.000000000001	1,000,000,000,000	12	
0.0000000000001	10,000,000,000,000	13	
0.00000000000001	100,000,000,000,000	14	

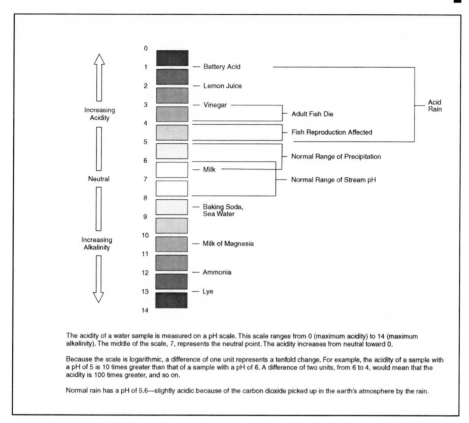

The acidity of a water sample is measured on a pH scale. This scale ranges from 0 (maximum acidity) to 14 (maximum alkalinity). The middle of the scale, 7, represents the neutral point. The acidity increases from neutral toward 0.

Because the scale is logarithmic, a difference of one unit represents a tenfold change. For example, the acidity of a sample with a pH of 5 is 10 times greater than that of a sample with a pH of 6. A difference of two units, from 6 to 4, would mean that the acidity is 100 times greater, and so on.

Normal rain has a pH of 5.6—slightly acidic because of the carbon dioxide picked up in the earth's atmosphere by the rain.

Figure 2-1 The pH scale

Figure 2-1 shows the range of the pH scale by identifying values for some common substances. As noted in the figure, "a difference of one pH unit represents a tenfold change." For example, lye with a pH of 13, is 10 times more alkaline than ammonia with a pH of 12. In a water treatment plant, a reading of pH 7 for the raw water and pH 7.5 for the finished water is not just a change of five increments; it would reflect a change from 7 to 35 in equal units. Many chemical reactions are either accelerated or slowed to a crawl by pH adjustments. For example, manganese (Mn^{+2}) oxidizes many times more slowly at pH 7 than it does at pH 8.5.

No single finished water pH applies as a standard for all plants. Finished water pH targets should be set by distribution system and consumer issues, including corrosion capabilities. Water with pH 8 may be an answer to a corrosion problem in one plant, while water with pH 7 will do the same in another plant. The Langelier Saturation index (LSI) and Aggressive index (AI), discussed in the following section, offer a starting point, but this discussion should not be the final word. An operator concerned about pH is well advised to consult a knowledgeable adviser.

Langelier and Aggressive Indexes

The Langelier Saturation index (LSI) measures a solution's ability to dissolve or deposit calcium carbonate, and it often serves as an indicator of the corrosivity of water. The index is not related directly to corrosion, but it is related to the formation of deposits of a calcium carbonate film or scale that can insulate pipes, boilers, and other system components from contact with water. This theory assumes even deposits, a situation not likely in the real world. When no protective scale is formed, water is considered to be aggressive, and corrosion can occur. Highly corrosive water can cause system failures or result in health problems from dissolved lead and other heavy metals. On the other hand, too much scale can damage water systems, coat adsorption media (such as manganese greensand and pyrolusite) with scale, or increase the sizes of filter media grains. Media coating is not common in potable water plants (although it occurs from time to time), but scale buildup is regularly seen in industrial boilers.

In developing the LSI, Langelier derived an equation for the pH at which water is saturated with calcium carbonate ($CaCO_3$), assigning this value the symbol pH_S. This equation is based on the equilibrium expression for calcium carbonate solubility and bicarbonate dissociation. To improve the approximation of actual conditions, pH_S calculations were modified to include the effects of temperature and ionic strength.

LSI is defined as the difference between actual pH (measured) and the calculated pH_S. The magnitude and sign (< or > 0) of the LSI value show a water's tendency to form or dissolve scale and thus inhibit or encourage corrosion.

A water treatment plant operator should exercise great care in applying information developed using the LSI. The *Hach Water Analysis Handbook* (1992) cautions, "As with L[S]I, the AI [Aggressive index, a similar measure] is not a quantitative measure of corrosion but is a general indicator of the tendency for corrosion to occur and should be used with proper reservation." Neither corrosion nor scale deposition are on–off chemical occurrences of much practical importance, but these processes may interest a plant operator if the water shows a strong tendency.

Current research efforts are examining the basic premises of the Langelier approach. Obtain up-to-date information to resolve any concern about corrosion. Although information obtained from the LSI is not quantitative, it can give useful estimates of water treatment requirements, and it can serve as a general indicator of the corrosivity of water.

The formula and necessary tables of values are included in the *Hach Water Analysis Handbook* (1992). Several other chemistry handbooks also contain the formula and tables. The calculation requires the following values as input: water temperature in °C, total dissolved solids (TDS) in mg/L, calcium hardness, total alkalinity in mg/L $CaCO_3$, and actual pH.

Chemical Forms of Iron and Manganese

Iron in its simplest soluble form, the form most often encountered in raw well water or raw impounded water, is ferrous iron. Manganese in it simplest soluble form is manganous manganese. Both have a valence (explained in the next section) of 2, i.e., Fe^{2+} and Mn^{2+}.

Fe and Mn in surface water are no different chemically from Fe and Mn in water from a well source. Local rain runoff into an impoundment (earthen reservoir) leaches Fe and Mn as it flows toward the impoundment; river water pumped or diverted into an impoundment carries trace amounts of Fe and Mn. Over time, levels of Fe and Mn build up on the bottom of the impoundment. Where inversions take place, as described in chapter 1, both Fe and Mn cycle from bottom to top and back to the bottom again.

Many treatment plants that draw from surface sources do not key on Fe and Mn removal, so they live with some residual levels during periods of impoundment inversions, usually for only several days twice a year. Surface water treatment plants that use lime softening have no difficulty in removing Fe and Mn; an efficient conventional plant (i.e., a coagulation/sedimentation/filtration plant) can usually remove Fe easily if a preoxidation process precedes filtration. Manganese removal is sometimes a challenge for conventional surface-water treatment plants, however. For example, Fe and Mn complexed by low-molecular-weight organics that may not be coagulated (as described later in this chapter) would pass right through a granular media filter. A discussion of molecular weights is too advanced for this handbook, but this possibility emphasizes that the location of the raw-water intake is critical in the control of Fe and Mn levels; an intake too close to the bottom of the impoundment collects high levels of Fe and Mn, while an intake too close to the surface collects high levels of algae.

Some Fe and Mn removal plants draw raw water from both wells and impoundments. Seasonal changes in Fe and Mn levels require additional monitoring and process adjustments.

Understanding Valence Numbers

Valence, touched upon earlier in the chapter, is the number of electrons that an atom or radical can lose, gain, or share with other atoms or radicals. *Valence* has the same meaning as *combining power*. One may say it is the relative worth of an atom of an element in combining with the atoms of other elements to form compounds. A positive or negative whole number ranging from 0 to 7 expresses the valence of an atom or radical.

If an atom loses electrons from its outermost shell, it has a positive valence. If an atom loses one electron from its outermost shell, it is said to have a valence of 1; similarly, if an atom loses two electrons from its outermost shell, it is said to have a valence of 2. If the outermost shell gains electrons, the atom has a negative valence. An atom that gains one electron has a valence of -1; one that gains two electrons has a valence of -2, and so on.

The manganese in $MnO_2(s)$, such as pyrolusite and the artificial coating on manganese greensand, has a valence of 4, expressed as Mn^{4+}. The manganese in potassium permanganate ($KMnO_4$) has a 7 valence, expressed as Mn^{7+}. In forming the chemical compound $KMnO_4$, the manganese atom lost (or gave up) seven electrons.

An element's valence number states the number of electrons of the element associated with formation of a particular compound. Since this number may differ for different compounds of an element, it follows that any element may have more than one valence number. For example, iron combines with oxygen to form two different compounds: FeO, a black oxide, and Fe_2O_3, a red oxide composed of two atoms of iron and three atoms of oxygen. The red oxide is commonly referred to as *rust*. The black oxide is frequently observed in water plants, particularly underneath the scale of corroded pipe. Iron also forms many compounds with other elements in which it may have a valence number of 2 or 3.

The reasons for some facts of chemical life are unknown. Manganese can have valence numbers 2, 3, 4, 6, and 7:

- Mn^0 is manganese metal.
- Mn^{2+}, pale pink in color, is the manganous ion Mn^{+2}.
- Mn^{3+}, red–violet in color, is the manganic ion Mn^{+3}.
- Mn^{4+}, brown–black in color, is the valence of the Mn in solid pyrolusite, $MnO_2(s)$.
- Mn^{6+}, dark bottle-green in color, is the manganate ion in MnO_4^{2-}.
- Mn^{7+}, intense purple in color, is the permanganate ion in MnO_4 (as in $KMnO_4$).

Why manganese can't have a 5 valence is a puzzler.

Compounds of Fe and Mn

Some waters contain many Fe and Mn compounds. The following compounds are sometimes found in groundwaters, although Fe and Mn compounds don't usually prevent removal of the metals: Fe and Mn hydroxides, $Fe(OH)_2$ and $Mn(OH)_2$; Fe and Mn bicarbonates, $Fe(HCO_3)_2$ and $Mn(HCO_3)_2$; Fe and Mn sulfates, $FeSO_4$ and $MnSO_4$; and others.

It is difficult to establish whether or not the Fe and Mn in a raw water is organically bound, as described later in the chapter, and, if so, in what percentage. Determination of organic complexing in a specific water is often just speculation. If organic carbons are found in the water to be treated, but Fe and Mn can be removed to low levels, then the levels of organic carbons do not affect the removal process and are of no concern, except as they may contribute to potential trihalomethane (THM) formation. However, difficulty in removing Fe and Mn to low levels using the preoxidation/sedimentation/filtration process suggests a need for investigation. The presence of more than 2.0 to 2.5 mg/L organic carbons and/or hydrogen sulfide and/or ammonia is sufficient reason to conduct a pilot study to determine the process needed to yield water meeting quality requirements.

Oxidation of Fe and Mn

The three most common forms of water treatment oxidants are oxygen (from the air), chlorine, and potassium permanganate. Other oxidants in use today include ozone, chlorine dioxide, ultraviolet light, and hydrogen peroxide. All have legitimate applications.

Iron in solution, as mentioned earlier, is known as *ferrous iron*, Fe^{2+}. Following oxidation it is ferric iron, Fe^{3+}. Manganese in solution is known as *manganous manganese*, Mn^{2+}. Following oxidation, it is manganic manganese, Mn^{4+}, although it may also take the form Mn^{3+}.

Oxidized Fe and Mn most often form iron hydroxide and manganic dioxide. Iron hydroxide can form a relatively large, sticky floc, which is very visible and easy to filter out of treated water. A coarse coal layer is often sufficient to filter out the iron hydroxide, though this is not true in every case. Some oxidized Fe precipitates require the addition of a polymer before removal. Manganic dioxide forms a far finer floc, so fine at times that a granular media filter will not remove it. The sticky iron hydroxide often traps or sticks to some of the oxidized manganese and aids in its removal.

Background on Organic Chemistry

This field is the specific branch of chemistry concerned with compounds of carbon. Carbon is unique among all other elements, not only because of its great reactivity, but also because of its ability to combine with itself in various ways to form a very large number of compounds. Many thousands of carbon compounds are known, and more are being discovered. The possibilities seem endless, especially as chemists design synthetic compounds for special purposes. All plant and animal tissue is composed of numerous carbon atoms in combination chiefly with hydrogen and oxygen and less commonly with nitrogen, phosphorus, sulfur, and metals.

Not all carbon compounds, by any means, are good for people, though. For example, some by-products resulting from combinations of chlorine and organic carbons are known as *trihalomethanes*, or THMs. Under laboratory conditions, THMs have caused cancer in animals. Trihalomethanes are therefore known as carcinogens. Obviously, every water treatment plant operator should recognize the dangers of producing chemical by-products harmful to humans. The old anaesthetic, chloroform, is trichloromethane ($CHCl_3$), one member of the THM grouping of undesirable by-products.

Carbon has a valence of +4. To understand this characteristic, picture the carbon atom as having four hooks. A hydrogen atom has one hook, oxygen has two hooks, nitrogen has three hooks, ferrous iron has two hooks, and ferric iron has three hooks. The concept gets more complicated with some other elements, like manganese, which has two-, three-, four, six-, and seven-hook variations. The "hooks" are called *bonds* or *valence*, as explained earlier in the chapter.

Organic Complexing of Fe and Mn

The whole question of organic complexing of Fe and Mn is difficult to address in a meaningful way for a small-plant operator. Studies of Fe and Mn organic complexing are few, no service offers on-site determination of organic complexing, and few laboratories are capable of identifying and quantifying troublesome organic acids. Still, recurring phenomena may display almost identical early warning signs, indicating at the very least the presence of organic compounds. Such a condition suggests a need to depart from preoxidation/sedimentation/filtration processes.

Experience with trouble calls from operators experiencing difficulty in removing Fe and Mn (especially manganese) below targeted levels has uncovered some common factors:

1. Well water is being drawn from shallow wells adjacent to flowing water, lakes, or sloughs.
2. A level of organic carbons over 2 mg/L is identified.
3. Some level of ammonia is quantified.
4. Some level of hydrogen sulfide is detected.

If the preoxidation/sedimentation/filtration process cannot remove Fe and Mn below target levels in a situation marked by these four warning flags, perhaps they can be removed to acceptable levels by treating them as organically complexed. If so, what practical need does an operator have to determine whether or not organic complexing is established beyond a shadow of a doubt? Also, does the operator need to know why the Fe and Mn are in a colloidal state or go to a colloidal state when an oxidant is applied? Few people would demand these explanations.

A treatment plant operator is expected to find a practical, safe, effective, affordable way of meeting Fe and Mn removal objectives. At the same time, the operator should have some knowledge of organic chemistry to allow identification of certain characteristics of the raw water being treated, without expending great amounts of time and money conducting exhaustive scientific determinations. At the very least, an operator needs to know the information presented in this section.

The most troublesome of organic compounds in Fe and Mn removal processes are the organic acids, which contain one or more carboxylic groups (COOH). In this carboxyl group, the hydrogen (H) is "active," i.e., it will ionize slightly to give acidic H^+ ions. More importantly, the active hydrogen will exchange for ions such as calcium, magnesium, sodium, iron, manganese, nickel, etc., exactly as occurs in zeolite softening. The common, naturally occurring organics encountered in water result from decomposition of organic materials (e.g., plants, leaves, and algae). A common organic compound is humic acid, which has the approximate formula:

$$C_{140}H_{126}O_5(COOH)_{17}(OH)_7(CO)_{10}OCH_3$$

Chemists have defined no exact formula for this acid. A molecule of humic acid contains about 17 carboxylic groups, so it shares some characteristics with a water-soluble, ion exchange resin.

Some ion exchange "zeolites," or more accurately resins, are water insoluble and contain thousands of carboxylic sites. Called *weak acid cation* exchange resins, they are used widely in the water industry.

Anionic polymers also contain the carboxyl as their active sites. As H ionizes to H^+, the rest of the carboxyl group has a negative charge.

This discussion relates to Fe and Mn removal because hydrogen of the carboxyl group will exchange with Fe^{+2} and Mn^{+2} ions, each of these soluble, two-valent ions exchanging with two hydrogens. This process is commonly described as *organic binding* or *complexing*.

An attempt to oxidize the bound Fe^{+2} or Mn^{+2} may or may not succeed. If oxidation occurs, creating insoluble ferric iron (Fe^{+3}) or changing Mn^{+2} to insoluble Mn^{+3} or Mn^{+4}, the iron or manganese may still be held by the ion exchange bonding. In such a case, the process may yield complexed, oxidized Fe/Mn forming a tiny organics/Fe/Mn clump or colloid, which carries an overall negative charge and is very difficult to remove, as it does not settle and passes through filters. A nonionic or anionic polymer may help to agglomerate colloids to form units large enough to settle out or become trapped in filters. Organic complexing can result from ion exchange, although not all chemical binding works this way.

Attempts to identify precisely the organics present in a raw water require extremely difficult and costly methods seldom justifiable as a routine control measure. Instead, operators monitor total organic carbon (TOC) as an indicator to warn of a potential for organic complexing. Total organic carbon levels greater than 2.0 to 2.5 mg/L indicate a potential removal problem using the preoxidation/direct filtration process.

In carbon compounds, hydrogen ions may have been exchanged for iron ions, creating difficulties in oxidizing the iron. When iron can be oxidized, it still may remain bound to its host carbon rather than agglomerating with other oxidized iron ions to form a particle big enough to be removed by conventional granular filters (i.e., the particles remain very fine or colloidal in size). Knocke et al. (1990, pp. 3, 5) reported:

> In many surface waters, a significant amount of the iron exists in a complexed form with organic matter. This complexation effect has been noted for decades; for example, Weston (1909) observed that certain waters containing organic matter were able to hold iron in solution for indefinite periods of time after aeration. Weston believed that organic compounds were involved in an attachment with the iron that prevented subsequent iron precipitation.

The impact of such complexation of iron oxidation by organic matter has been reported by several authors. For example, Hem (1960) demonstrated that tannic acid had the ability to form complexes with ferrous iron and significantly retard its subsequent oxidation by oxygen. Jobin and Ghosh (1972) showed that both humic acids and tannic acids have the ability to complex iron and retard its oxidation upon exposure to O_2(aq), allowing essentially no iron oxidation to take place.

Theis and Singer (1973, 1974) presented excellent research results aimed at elucidating the ability of various organic compounds to complex iron and retard its removal by contact with O_2(aq). The authors showed that tannic acid, gallic acid, pyrogallol, and other products of natural vegetative decay can effectively retard the oxidation of Fe(II) for several days, even in waters saturated with O_2(aq).

Fe and Mn that are organically complexed can usually be oxidized using chlorine or $KMnO_4$, given the appropriate dosage, pH, and detention time. Organically complexed iron can sometimes be oxidized by oxygen from simple aeration. In this state, its size is usually very fine or colloidal, so it is too small to be removed by a granular media filter. One removal option is to neutralize the surface charge using coagulants like alum, iron salts, poly aluminum compounds, or cationic polymers, followed by settling and/or direct filtration.

Oxidized Fe and Mn species not organically complexed, and their compounds, can remain colloidal following oxidation. Again, the removal method described in the last paragraph is one option. In some cases, however, the raw water contains elements and compounds that interfere with the action of certain coagulants.

Although every treatment plant operator should display scientific curiosity, the complexation of manganese (or lack of it) is well beyond the research capabilities of operators with limited training and inadequate laboratory equipment. Priorities focus, not on confirming manganese/organic complexing, but on finding the process that will remove the manganese. Analysis should determine how the raw-water manganese behaves, but scientific terminology for defining that behavior is not important to the operator. With these limitations in mind, one basis for organic complexing is ion exchange.

Adsorption Removal Methods

In most cases, an adsorption process works well for the removal of manganese. For optimum chemical cost and simplicity of operations, use of the adsorption process often makes the most sense, especially in small plants. Pyrolusite, manganese greensand, or any medium coated with MnO_2(s) has the capacity to adsorb Mn^{2+}. Regeneration is accomplished by exposing the medium to a chlorine solution or a solution containing $KMnO_4$. The amount of regenerant can be calculated based on the volume of water treated, amount of Fe and Mn in the raw water (milligrams per litre), and theoretical oxidant demand.

The major risk in using an adsorption process to remove iron results from gradual coating of the $MnO_2(s)$ surfaces with oxidized iron, which can result in media blinding. To prevent or minimize this problem, a pilot test is usually undertaken to evaluate a system to oxidize the iron and remove it by filtration in a top coal layer. The Mn is then removed by adsorption. Iron oxidation/filtration combined with manganese adsorption is a dual process practiced successfully in many small plants.

Chapter 3
Physics and Mathematics of Iron and Manganese Removal

Hydraulics and Applicable Water Physics

Hydraulics is the name given to the branch of science concerned with fluids at rest and in motion. *Hydrostatics* refers to fluids at rest, and *hydrodynamics* refers to fluids in motion. The fluid dealt with in this handbook is water, but many of the same principles apply to air and individual gases.

Quantitative descriptions of hydraulic events can become very complicated, using many coefficients, rates, mathematical descriptions of friction, flow curves, and diverse other factors. This part of the discussion will cover only the basics of hydraulic heads and the effects of heads and pressure differentials on certain granular filter media.

Types of Heads

Head is the amount of energy exerted by a unit quantity of water at its given location. Ordinarily, the energy is expressed in metre-kilograms (m-kg)(pound feet), and the unit quantity of water considered is 1 kg (2.2 lb). The head, then, is expressed in metre-kilograms of energy per kilogram of water, or

$$\frac{m \times kg}{kg} = m \qquad (3\text{-}1)$$

For this reason, all heads can be expressed in metres. Energy in water can result from its elevation, pressure, or velocity. These energies are called *static (elevation) head*, *pressure head*, and *velocity head*. Other heads are pump head, which equals the metre-kilograms of energy given to each kilogram of water passing through a pump, and friction head, which is the energy lost due to friction within the fluid and against the walls of a pipe or channel.

Static (or elevation) head. Elevations are expressed as the vertical distance from some base level (known point or reference plane), such as sea level, the surface of the ground, the bottom of a filter, or some other arbitrarily chosen level. See Figure 3-1.

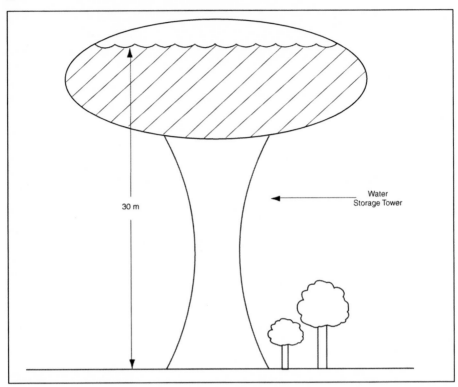

Figure 3-1 Static head

For example, water that is 3 m (9.8 ft) above the bottom of a filter exerts 3 m-kg of energy, and its static (or elevation) head is 3 m (9.8 ft).

Pressure head. Pressures are expressed in terms of force per unit area, such as kilopascals (kPa) or pounds per square inch (psi). So, 1 kPa is the unit of force that will give a mass of 1 kg of water an acceleration of 1 metre per second per second (m/s^2). If water could be piled up to exert a pressure of 10 kPa (1.5 psi), the pressure head would be 1 m, since all heads can be expressed in metres.

Conversion. Some common conversion formulas follow; refer to appendix A for other conversions.

- To convert psi to kPa, multiply psi by 6.895.
- To convert kPa to psi, multiply kPa by 0.14504.
- To convert lb/ft^2 (pounds per square foot) to kPa, multiply lb/ft^2 by 0.0479.

Pressure Differential and Head Loss

Many plants remove Fe and Mn using pressure filters with manganese greensand as the principle filtering medium. Manganese greensand is a glauconite sand coated with manganese dioxide, $MnO_2(s)$. At pressure differentials exceeding 55.2 kPa

(8 psi) the MnO_2(s) coating can undergo fracturing that could reduce the medium's adsorptive/exchange capacity for Fe and Mn. Repeated exposure to pressure differentials exceeding 55.2 kPa (8 psi) could also result in the breakdown of some of the base glauconite particles.

Figure 3-2 shows the top gauge reading 69.0 kPa (10 psi) and the bottom gauge reading 13.8 kPa (2 psi). The pressure drop (or differential), therefore, is 55.2 kPa (8 psi).

Applying the information on static and pressure head presented earlier to Figure 3-2, the distance from the top gauge to the bottom gauge equals 244 cm (8 ft). The static head is 244 cm, or 2.44 m (8 ft), but the pressure head, referred to as P_s, is 244 (8) divided by 10 (2.3), as calculated earlier, or 24 kPa (3.5 psi).

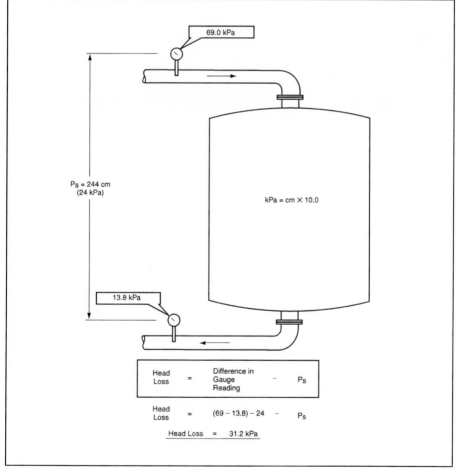

Source: Antratech Western Inc.

Figure 3-2 Head loss example

The formula for calculating head loss in kPa is:

$$\begin{aligned} \text{head loss} \quad &= \quad \text{difference in gauge readings minus } P_s \\ &= \quad (69 - 13.8) \text{ kPa} - 24 \text{ kPa} \\ &= \quad 55.2 - 24 \\ &= \quad 31.2 \text{ kPa} \end{aligned} \qquad (3\text{-}2)$$

Manganese greensand performs in the same way regardless of the source of pressure. If a filter bed is plugged with particles (manganese, iron, silt, fines, etc.), pressure builds up regardless of the type of filter (closed pressure filter or open gravity filter). If the pressure differential builds to a point that it exceeds the strength of the particles, the glauconite and its $MnO_2(s)$ coating could be fractured, eliminating some $MnO_2(s)$ coating and ultimately reducing the bed's adsorptive capacity. The manufacturer of manganese greensand specifies a maximum pressure differential of 69 kPa (10 psi); some authors of this book have observed particle/coating fracturing at 55 kPa (8 psi). Although documentation of this phenomenon could not be found, the operator is well-advised to heed these maximum pressure differential figures.

Figure 3-2 can be a very misleading introduction to head loss. The formula and calculation are correct, assuming that no water is flowing. However, if water were flowing through the lines and filter, depending on the amount of resistance of the filter bed and the underdrain, the static head (shown as P_s) might be only 60 to 120 cm (24 to 48 in.), a calculation best left to someone familiar with the process. Many other factors can affect an actual head loss calculation. If a pump located in the effluent line is actually pulling water through the filter, that's another factor to consider, and results could vary with others.

In many water treatment plants (WTPs), automatic devices trigger backwash cycles based on head loss. A number of different devices can be used for this purpose, but how does information about head loss help a plant operator? The operator simply needs to recognize the big difference between head loss and pressure differential. Head loss is very important to a plant designer, while pressure differential is important to the operator. If, for example, an operator sees a pressure differential near 55 kPa (8 psi) on a manganese greensand filter, this condition can signal the need for a backwash, regardless of any other sensing equipment on the filter.

Flow Rates

Flow rates can be expressed in metres per hour (m/h); in filtration, m/h is actually an expression for cubic metres per square metre per hour ($m^3/m^2/h$).

Most often, flow rates for both filtration and backwash are expressed in gallons per minute per square foot (gpm/ft^2). In this system, a filtration flow rate of 2–3 gpm/ft^2 is considered appropriate for a manganese greensand pressure or gravity filter. Greensand backwashing requires 10–12 gpm/ft^2 depending on water temperature. To convert gpm/ft^2 to m/h, multiply gpm/ft^2 by 2.4476.

Water treatment plant gauges often record mixtures of pounds per square inch, feet of water, and kilopascals. Training manuals and reference books use either US customary measurements or a combination of customary and metric units. For practical functioning, an operator should keep a metric converter pocket calculator handy, along with the common conversion factor tables at the back of this handbook.

Mathematics

Calculating Chemical Dosages

Optimum chemical dosages can critically affect any Fe and Mn removal process. As an example, manganous manganese is the simplest form of soluble reduced manganese; its symbol is Mn^{2+}. When fully oxidized, it changes to manganic manganese, with the symbol Mn^{4+}. In order to reach that 4+ valence, each mg/L of manganese requires 1.3 mg/L of chlorine (in the form of hypochlorous acid, HOCl), or 1.92 mg/L of potassium permanganate ($KMnO_4$). Actual chemical oxidant demand of a raw water often exceeds the theoretical demand for treatment of Fe and Mn alone, because other constituents, such as organic carbons and ammonia, contribute to oxidant demand, as well.

The most commonly used oxidants in water treatment applications include
- air (atmospheric oxygen) as O_2 aq (meaning oxygen in a water solution, or aqueous oxygen)
- ozone as O_3 aq
- chlorine (Cl_2) in the form of hypochlorous acid as HOCl
- chlorine dioxide as ClO_2
- potassium permanganate as $KMnO_4$

Table 3-1 details the reactions that result.

The amount of chemical required depends on the actual form and strength of the product used. For example, chlorine gas is 100 percent chlorine, calcium hypochlorite is 65% available chlorine, and commercial sodium hypochlorite is 12 percent available chlorine. (Ordinary household bleach is about 5–6 percent available chlorine.) Available chlorine in calcium and sodium hypochlorite is taken as the percentage of OCl present. Depending on the brand name, potassium permanganate is 97–99 percent strength. Potassium permanganate normally comes as solid crystals, calcium hypochlorite in granular or pellet form, and sodium hypochlorite in liquid form. Both solid products require mixing with water to make a solution, which is then fed into the process flow. Sodium hypochlorite liquid can be fed neat (i.e., full strength as it comes from the supplier) or diluted with water.

Caution: When preparing solutions of water and sodium hypochlorite, the two-tank method should always be used to avoid the potential for plugging chemical pumps or chemical feed lines with precipitates. Following batch mixing, the solution should be allowed to settle and stabilize. The stable solution should then be drawn off into a second container into which the chemical pump's suction end is placed.

Table 3-1 Theoretical reaction stoichiometry, Fe^{2+} and Mn^{2+}

Metal/Oxidant	Reaction	Stoichiometry mg/L Oxidant Versus Metal
Iron		
O_2 (aq)	$2Fe^{2+} + {}^1\!/_2O_2 + 5H_2O \rightarrow 2Fe(OH)_3(s) + 4H^+$	0.14 mg : 1 mg Fe
O_3 (aq)	$2Fe^{2+} + O_3 + 5H_2O \rightarrow 2Fe(OH)_3(s) + O_2 + 4H^+$	0.43 mg : 1 mg Fe
HOCl	$2Fe^{2+} + HOCl + 5H_2O \rightarrow 2Fe(OH)_3(s) + Cl^- \, 5H^+$	0.64 mg : 1 mg Fe
$KMnO_4$	$3Fe^{2+} + MnO_4 + 7H_2O \rightarrow 3Fe(OH)_3(s) + MnO_2 + 5H^+$	0.94 mg : 1 mg Fe
ClO_2	$5Fe^{2+} + ClO_2 + 13H_2O \rightarrow 5Fe(OH)_3(s) + Cl^- + 11H^+$	0.24 mg : 1 mg Fe
Manganese		
O_2 (aq)	$Mn^{2+} + {}^1\!/_2O_2 + H_2O \rightarrow MnO_2(s) + 2H^+$	0.29 mg : 1 mg Mn
O_3 (aq)	$Mn^{2+} + O_3 + H_2O \rightarrow MnO_2(s) + O_2 + 2H^+$	0.88 mg : 1 mg Mn
HOCl	$Mn^{2+} + HOCl + H_2O \rightarrow MnO_2(s) + Cl^- + 3H^+$	1.30 mg : 1 mg Mn
$KMnO_4$	$3Mn^{2+} + 2KMnO_4 + 2H_2O \rightarrow 5MnO_2(s) + 4H^+$	1.92 mg : 1 mg Mn
ClO_2	$Mn^{2+} + 2ClO_2 + 2H_2O \rightarrow MnO_2(s) + 2ClO^{2-} + 4H^+$	2.45 mg : 1 mg Mn

The combination of water and sodium hypochlorite forms caustic, which reacts chemically with calcium and magnesium, producing softened water and precipitates. The volume of precipitates can be substantial enough to be a very bothersome problem. The choice whether or not to use treated water from the clear well in making the solutions doesn't matter much when comparing hardness that may reach 400 mg/L to 2–5 mg/L iron and less than 1 mg/L manganese.

Depending on the quantity of solution required daily and the chemical used to make it, solution strengths typically range from 0.5 to 4 percent. Potassium permanganate does not dissolve easily or rapidly in cold water, so solution strengths should not exceed 4 percent when using cold water (see Figure 5-2 later in the handbook). Also, the mixer should run for 30 minutes or more.

Solution strength in percentage is equivalent to the weight of the chemical divided by the weight of the water in which it is mixed, assuming the chemical product is made up 100 percent of the active ingredient. If not, the following formula is used:

$$\text{percentage solution strength} = \frac{\text{weight of chemical added} \times \text{percentage available active ingredient}}{\text{weight of water in solution}} \quad (3\text{-}3)$$

Using Eq 3-3, find the solution strength produced by adding 1,000 grams of $KMnO_4$ to 50 L of water.

$$\text{percentage solution strength} = \frac{1{,}000 \text{ g KMnO}_4 \times 99 \text{ percent*}}{50{,}000 \text{ g (same as mL) water}}$$

$$\text{solution strength} = 1.98 \text{ percent}$$

Equivalents:
- When weighing water in metric units, 1 L = 1,000 g
- In Imperial units, 1 gal = 10 lb
- In US customary units, 1 gal = 8.34 lb

To make a solution of a specified strength, calculate the weight of the chemical required as:

$$\text{weight of chemical} = \frac{\text{percentage strength} \times \text{weight of water}}{\text{percentage active ingredient}} \qquad (3\text{-}4)$$

Example 1: To prepare 75 L of 2 percent $KMnO_4$ solution, the calculation is:

$$KMnO_4 = \frac{2\% \times 75 \times 1{,}000}{99\%^{\dagger}}$$

$$= 1{,}515 \text{ g} = 1.52 \text{ kg}$$

Example 2: To prepare 45 Imperial gal of 3 percent $KMnO_4$ solution, the calculation is:

$$KMnO_4 = \frac{3 \times 45 \times 10}{99\%}$$

$$= 13.64 \text{ lb, or } 13 \text{ lb } 10 \text{ oz}$$

The information needed to calculate the amount of chemical required includes the desired concentration, the amount of water to be treated, and the solution strength of the chemical product. The formulas in the remainder of this chapter are used for various dosage and concentration calculations.

1. Dosage: How Much Chemical Is Needed?

$$\begin{array}{l}\text{weight} \\ \text{of chemical}\end{array} = \frac{\begin{array}{c}\text{dosage} \\ \text{(in mg/L)} \times \text{weight of water treated} \times 100\%\end{array}}{1{,}000{,}000 \times \text{percentage active ingredient}} \qquad (3\text{-}5)$$

---

*Potassium permanganate active ingredient expressed as a percentage.

†The $KMnO_4$ product has 99 percent active ingredient in this example.

Same formula in metric units:

$$\text{grams of chemical} = \frac{\text{dosage (in mg/L)} \times \text{litres treated} \times 1{,}000 \times 100\%}{1{,}000{,}000 \times \text{percentage active ingredient}} \quad (3\text{-}6)$$

Same formula in Imperial units:

$$\text{pounds of chemical} = \frac{\text{dosage (in mg/L)} \times \text{gallons treated} \times 10 \times 100\%}{1{,}000{,}000 \times \text{percentage active ingredient}} \quad (3\text{-}7)$$

To calculate the volume of a treatment solution, use the following formula for both metric and Imperial units:

$$\text{volume of solution} = \frac{\text{weight of chemical required} \times \text{percentage active ingredient}}{\text{percentage solution strength} \times \text{weight of water (litres or gallons)}} \quad (3\text{-}8)$$

2. Dosage: How Many Milligrams per Litre Need to Be Fed?

$$\text{dosage (in mg/L)} = \frac{\text{weight of chemical used} \times 1{,}000{,}000 \times \text{percentage active ingredient}}{\text{weight of water treated} \times 100\%} \quad (3\text{-}9)$$

Same formula in metric units:

$$\text{dosage (in mg/L)} = \frac{\text{grams of chemical used} \times 1{,}000{,}000 \times \text{percentage active ingredient}}{\text{litres treated} \times 1{,}000 \times 100\%} \quad (3\text{-}10)$$

For solutions:

$$\text{dosage (in mg/L)} = \frac{\text{percentage solution strength} \times \text{litres of solution used} \times 1{,}000{,}000}{\text{litres treated} \times 100\%} \quad (3\text{-}11)$$

Same formula in Imperial units:

$$\underset{\text{(in mg/L)}}{\text{dosage}} = \frac{\overset{\text{pounds of}}{\text{chemical used}} \times 1{,}000{,}000 \times \overset{\text{percentage}}{\text{active ingredient}}}{\text{gallons treated} \times 10 \times 100\%} \quad (3\text{-}12)$$

For solutions:

$$\underset{\text{(in mg/L)}}{\text{dosage}} = \frac{\overset{\text{percentage}}{\text{solution strength}} \times \overset{\text{gallons of}}{\text{solution used}} \times 1{,}000{,}000}{\text{gallons treated} \times 100\%} \quad (3\text{-}13)$$

Example Dosage Calculations: Metric
Assumptions:
Treatment handles 450,000 L per day.
Chlorine demand is 7.5 mg/L.
Potassium permanganate demand is 1.5 mg/L.
Actual daily usage is 92 L of 3.5 percent calcium hypochlorite solution,
and 33 L of 1.75 percent $KMnO_4$ solution.

Questions:
1. How much sodium hypochlorite is needed each day?
2. How much potassium permanganate is needed each day?
3. If calcium hypochlorite is mixed to a 3.5 percent chlorine solution, how much would be used in a day?
4. If potassium permanganate is mixed to a 1.75 percent solution, how much would be used in a day?
5. What average Cl_2 and $KMnO_4$ dosages result?

Answers:

Question 1

$$\underset{\text{of chemical}}{\text{grams}} = \frac{\text{dosage} \times \text{litres} \times 1{,}000 \times 100}{1{,}000{,}000 \times \text{percentage active ingredient}}$$

$$= \frac{7.5 \times 450{,}000 \times 1{,}000 \times 100}{1{,}000{,}000 \times 12}$$

$$= \frac{7.5 \times 450 \times 100}{12}$$

$$= 28{,}125$$

Remember, 1 L equals 1,000 g, so 28,125 g or 28.125 L of sodium hypochlorite is needed.

Question 2

$$\text{grams} = \frac{1.5 \times 450{,}000 \times 1{,}000 \times 100}{1{,}000{,}000 \times 99}$$

$$= \frac{67{,}500}{99}$$

$$= 681.8 \text{ g}$$

Since 1,000 g equals 1 kg, 682 g equals about 0.7 kg, the amount of $KMnO_4$ needed each day.

Question 3 The first half of this two-part problem is the same as question 2.

$$\text{grams} = \frac{7.5 \times 450{,}000 \times 1{,}000 \times 100}{1{,}000{,}000 \times 65}$$

$$= 5{,}192 \text{ g} = 5.2 \text{ kg calcium hypochlorite needed}$$

The second half of question 3 asks:

$$\text{litres} = \frac{\text{grams of chemical} \times \text{percentage active ingredient}}{\text{percentage solution strength} \times 1{,}000}$$

$$= \frac{5{,}192 \times 65}{3.5 \times 1{,}000} = \frac{337{,}480}{3{,}500}$$

$$= 96.4 \text{ L used daily}$$

Question 4

$$\text{litres} = \frac{682 \times 99}{1.75 \times 100} = \frac{67{,}518}{1{,}750}$$

$$= 38.58, \text{ or about } 39 \text{ L used daily}$$

Question 5

$$\text{Cl}_2 \text{ dosage} = \frac{\text{percentage solution strength} \times \text{litres of solution} \times 1{,}000{,}000}{\text{litres treated} \times 100}$$

$$= \frac{3.5 \times 92 \times 1{,}000{,}000}{450{,}000 \times 100}$$

$$= 7.2 \text{ mg/L}$$

$$\text{KMnO}_4 \text{ dosage} = \frac{1.75 \times 33 \times 1{,}000{,}000}{450{,}000 \times 100}$$

$$= 1.3 \text{ mg/L}$$

Metric Formula Summary

1. chemical required = chemical demand + residual (or excess)

To calculate:

2. Solution strength

$$\text{percentage strength} = \frac{\text{kilograms of chemical added} \times \text{percentage active ingredient}}{\text{litres of water in solution} \times 1{,}000}$$

3. Chemical needed for a dilute solution

$$\frac{\text{grams of chemical}}{} = \frac{\text{percentage strength} \times \text{litres of water in solution} \times 1{,}000}{\text{percentage active ingredient}}$$

4. Chemical needed for a particular dosage

$$\frac{\text{grams of chemical}}{} = \frac{\text{dosage} \times \text{litres treated} \times 1{,}000 \times 100\%}{1{,}000{,}000 \times \text{percentage active ingredient}}$$

5. Volume of solution to be fed

$$\frac{\text{litres of solution}}{} = \frac{\text{grams of chemical needed} \times \text{percentage active ingredient}}{\text{percentage solution strength} \times 1{,}000}$$

6. Dosage applied—dry chemical feed

$$\text{dosage (mg/L)} = \frac{\text{grams of chemical} \times 1{,}000{,}000 \times \text{percentage active ingredient}}{\text{litres treated} \times 1{,}000 \times 100\%}$$

7. Dosage applied—solution feed*

$$\text{dosage (mg/L)} = \frac{\text{percentage solution strength} \times \text{litres of solution used} \times 1{,}000{,}000}{\text{litres treated} \times 1{,}000 \times 100\%}$$

Imperial Formula Summary

1. chemical required = chemical demand + residual (or excess)

To calculate:

2. Solution strength

$$\frac{\text{percentage solution strength}}{} = \frac{\text{pounds of chemical added} \times \text{percentage active ingredient}}{\text{weight of water in solution}}$$

*This formula applies to dilute solutions of about 5 percent or lower strength.

3. Chemical needed for a dilute solution

$$\frac{\text{pounds of}}{\text{chemical}} = \frac{\text{percentage strength} \times \text{weight of water in solution}}{\text{percentage active ingredient}}$$

4. Chemical needed for a particular dosage

$$\frac{\text{pounds of}}{\text{chemical}} = \frac{\text{dosage} \times \text{gallons treated} \times 10 \times 100\%}{1,000,000 \times \text{percentage active ingredient}}$$

5. Volume of solution to be fed

$$\frac{\text{gallons of}}{\text{solution}} = \frac{\text{pounds of chemical needed} \times \text{percentage active ingredient}}{\text{percentage solution strength} \times 10}$$

6. Dosage applied—dry chemical feed

$$\text{dosage (lb/gal)} = \frac{\text{pounds of chemical used} \times 1,000,000 \times \text{percentage active ingredient}}{\text{gallons treated} \times 10 \times 100\%}$$

7. Dosage applied—solution feed*

$$\text{dosage (lb/gal)} = \frac{\text{percentage solution strength} \times \text{gallons of solution used} \times 1,000,000}{\text{gallons treated} \times 100\%}$$

Practical Math

Detention and fill time calculations. Removal of Fe and Mn almost always requires the addition of an oxidant to the raw water to bring about a chemical transformation of the iron and manganese in solution to a filterable form. Raw-water pH and temperature as well as chemicals or compounds naturally present in the raw water determine the time needed to complete oxidation and formation of a filterable floc.

Detention time is the time required for a particular flow to travel through a tank. Another way to think of detention time is the time required for a given flow to fill a tank. The tank in this definition may be an actual tank designed to hold water over time (a detention or holding tank or vessel); it may also include part of a pressure filter vessel above the filter media, the piping from the dosing point to the filter entry, or a combination of all three. Typically, *detention time* as used in chemical pretreatment is equivalent to *contact time*, not to be confused with the C x T value, which describes the residual concentration of a disinfectant over time.

---

*This formula applies to dilute solutions of about 5 percent or lower strength.

Two variables affect detention time: the flow rate through the tank and the volume of the tank. Use this formula to determine detention time:

$$\text{detention time} = \frac{\text{volume of tank}}{\text{flow rate}}$$

Two other considerations influence detention time calculations.
1. Flow rate and volume must be expressed in the same terms. If the tank's volume is stated in cubic metres, then the flow rate must be stated in cubic metres per unit of time. Likewise, a volume in gallons requires a flow rate in gallons per unit of time (i.e., gpm, gph, gpd, and so forth). (3-14)
2. The time frame of the flow rate determines the time frame of the detention time calculation. For example, if the flow rate is stated as cubic metres per hour, the detention time is in hours; if the flow rate is stated in gallons per minute, the detention time is in minutes.

Example calculation:
A detention tank is 3 m long, 2 m wide, and 2.5 m deep to the full line. If the flow rate is 30 m^3/h, what is the detention time? First, the standard geometric formula (l x w x d) gives a tank volume of 15 m^3.

$$\text{detention time} = 15/30, \text{ or } 0.5 \text{ h } (30 \text{ min})$$

Calculating media surface area. For a square or rectangular filter, surface area is the product of the length multiplied by the width:

$$A \text{ (area)} = l \text{ (length)} \times w \text{ (width)}$$

In the previous example, the area equals 3 × 2, or 6 m^2.

For a round filter, apply the formula:

$$A = \text{pi} \times r \text{ squared, or } A = \pi r^2$$

The symbol for pi is π (π = 22 divided by 7, or 3.142; r refers to radius, i.e., one-half the diameter of a circle).

Example:
What is the surface area of the media in a cylindrical, vertical pressure filter with a diameter of 1.83 m (6 ft)? Since radius is half the diameter, r = 1.83/2 = 0.915; r^2 = $r \times r$ = 0.915 × 0.915 = 0.837. Therefore,

$$A = 3.142 \times 0.837$$
$$= 2.630 \text{ m}^2$$

In Imperial measurements, the calculation would be:

$$
\begin{aligned}
A &= 3.142 \times (3 \text{ ft} \times 3 \text{ ft}) \\
&= 3.142 \times 9 \text{ ft}^2 \\
&= 28.278 \text{ ft}^2
\end{aligned}
$$

Rounded to the first decimal place, the surface area is 28.3 ft^2.

Flow rate calculation. The round filter receives unfiltered water at the rate of 16 m^3 (cubic metres) per hour. What is the rate of filtration per square metre of filter surface area?

$$
\text{filtration rate} = \frac{\text{flow rate}}{\text{area}} \qquad \text{(3-15)}
$$

The example's filtration rate equals 16 m^3/h divided by 2.63 m^2, or 6.08 m^3/h/m^2, which reduces to 6.1 m/h.

In Imperial measurements, the calculation would begin by converting 16 m^3 to gallons; multiplying by 220 gives 3,520 Imp gph. Since filtration rates in Imperial units are usually given in gallons per minute, divide 3,520 by 60 to get 58.67 Imp gpm.

$$
\text{filtration rate} = 58.67/28.3 = 2.07, \text{ or } 2.1 \text{ Imp gpm/ft}^2
$$

In US measurements, the calculation would first, convert 16 m^3 to gallons; multiplying by 264.2 gives 4,227.2 gph divided by 60 equals 70.453 gpm.

$$
\text{filtration rate} = 70.453/28.3 = 2.49, \text{ or about } 2.5 \text{ gpm/ft}^2
$$

An operator unfamiliar with conversions of metric values to Imperial or US customary units should refer to the conversion factors available in appendix A.

Chapter 4

Microbiology of Iron and Manganese Removal

---∎

"Groundwater is an important source of water, especially in rural areas. This resource is of increasing concern to microbiologists" (Prescott, Harley, and Klein 1990).

The science of biology is concerned with the study of living organisms, including their habits, food requirements, and general life processes. The living organisms covered in this chapter are too small to be seen with the unaided eye; some can be seen with light microscopes that magnify up to 2,000 diameters. Hence, the name microbiology.

Every water treatment plant operator needs to monitor the impact of the common forms of groundwater bacteria on water treatment and finished water quality. This impact highlights the importance of adequate disinfection in the plant's clear well and distribution system.

Rust bubbles of corrosion appear on pipes, for example, due to microbially induced corrosion by specific bacteria. Certain bacteria form slimes that bioaccumulate (hold onto and build up) iron and manganese. These slimes are often evident in toilet tanks in areas where Fe and Mn abound in the raw water and those where raw well water does not undergo any treatment before use. These same slimes have been known to foul well pumps to the point that they force shutdowns. Other bacteria, known as *sulfur bacteria*, can generate unpleasant tastes and odors in finished water already in the clear well, reservoir, or distribution system. Some bacteria types are much more aggressive than others, perhaps even causing unwanted events in less than a day after their introduction.

An operator needs to know that active, growing, and dying organisms generate much of the material seen accumulating on pipes, filter walls, and clear well ladders, as well as that floating on the surface of the clear well and moving in distribution pipes.

Potential Effects of Microbiological Organisms

According to Cullimore (1993, pp. 12–13),

> Microbial events can include corrosion initiation, bioimpedance of hydraulic or gaseous flows (e.g., plugging, clogging), bioaccumulation of chemicals (e.g., localized concentration of heavy metals, hydrocarbons and/or radionu-

clides within a groundwater system), biodegradation (e.g., catabolism of potentially harmful organic compounds), biogenesis of gases (including methane, hydrogen, carbon dioxide, or nitrogen, which can lead to such events as the fracturing of clays, displacement of water tables, and differential movement of soil particles), and water retention within biofilms (which, in turn, can influence the rates of desiccation and/or freezing of solids). These events occur naturally within the biosphere as microbiologically driven functions within and upon the crust of the planet.

This paragraph introduces many uncommon terms and big words, but it does reinforce the point that microbiological activity can have major influences on groundwater sources and raw-water quality. As the operator gains knowledge of the water to be treated, the likelihood increases of choosing the appropriate treatment process. Figure 4-1 illustrates some of the events described in the quoted paragraph and in the following definitions:

- **Microbially induced fouling** (MIF)   This term refers to rust bubbles, layer after layer of corrosion inside piping, and buildups of microbiological slimes, along with any other accumulation that reduces the flow volume. In other words, it describes events that result in gradual plugging.

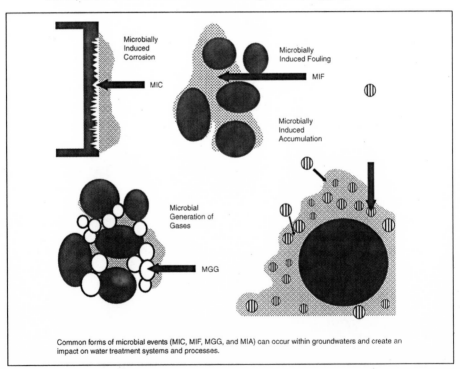

Common forms of microbial events (MIC, MIF, MGG, and MIA) can occur within groundwaters and create an impact on water treatment systems and processes.

Source: Reprinted with permission from D.R. Cullimore. 1993. Practical Manual of Groundwater Microbiology. Chelsea, Mich.: Lewis. Copyright CRC Press, Boca Raton, Florida.

Figure 4-1 Microbes in action

- **Microbially induced accumulation** (MIA)   As MIF takes place, a bioaccumulation also concentrates various ions, such as metallic elements (iron, manganese, etc.) in the form of dissolved or insoluble salts and organic complexes.
- **Microbial generation of gases** (MGG)   As groups of microorganisms mature in primarily anaerobic conditions, gases are likely to form. Corrosion bubbles, for example, can be filled with carbon dioxide, methane, hydrogen, and nitrogen. Such gases in their dissolved states can be parts of other slime groups. Under the right conditions, sufficient volumes of gases can form within biological growths to expand their sizes and reduce the diameter of a pipe, thus reducing the volume of water that the pipe can carry, despite the porous appearance of biological growth.
- **Microbially induced corrosion** (MIC)   Commonly called *rusting*, this process occurs when microbes attach themselves to a surface, for example, the wall of a steel pipe. They cover themselves up, creating anaerobic environments. Within their self-built homes, they may generate hydrogen sulfide (which promotes electrolytic corrosion) and/or organic acids (which can transform metals into solution). The results appear as surface pitting and/or rusting out (corrosion that extends through a metal wall of a pipe, filter, or filter false bottom).
- **Microbially induced relocation** (MIR)   As microorganisms live their lives, they consume nutrients and use oxygen until a change in the oxidation–reduction potential takes place. Under the right conditions, flowing water begins to rip apart the biofilm. Small pieces (or colonies) find new locations where nutrients and oxygen are available, and new colonies are created and grow until they too get ripped apart and flow to new homes.

The Importance of Oxidation–Reduction (redox)

As Figure 4-2 illustrates, much of the microbial growth that affects groundwater sources, and hence treatment of those groundwaters for removal of Fe and Mn, takes place in the zone where aerobic and anaerobic conditions meet.

According to Cullimore (1993, pp. 17–18),

> Redox has a major controlling impact on microorganisms through the oxidation-reduction state of the environment. Generally, an oxidative state (+Eh values) will support aerobic microbial activities while a reductive state (–Eh values) will encourage anaerobic activities which, in general, may produce a slower rate of biomass generation, a downward shift in pH where organic acids are produced, and a greater production potential for gas generation (e.g., methane, hydrogen sulfide, nitrogen, hydrogen, carbon dioxide). Aerobic microbial activities forming sessile growths are often noted to occur most extensively over the transitional redox fringe of –50 to +150 Mv.

(The glossary provides a definition of *sessile*.)

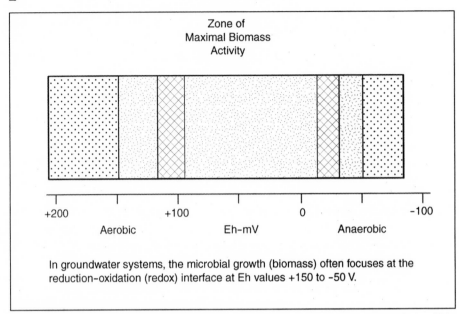

In groundwater systems, the microbial growth (biomass) often focuses at the reduction–oxidation (redox) interface at Eh values +150 to -50 V.

Source: Reprinted with permission from D.R. Cullimore. 1993. Practical Manual of Groundwater Microbiology. Chelsea, Mich.: Lewis. Copyright CRC Press, Boca Raton, Florida.

Figure 4-2 Where aerobic and anaerobic meet

Effect of Microbiological Activity on Fe and Mn Levels

When microbiological organisms (i.e., biofilms) join together to form a colony by attaching themselves to a surface, they simply grab their food from the passing water. (They may also use iron from certain pipe materials under some circumstances.) Obviously, any increase in the nutrient (i.e., food) level in the passing water speeds the colony's growth to maturity. At the same time, the biofilm can draw into itself such metals as iron, manganese, aluminum, copper, and zinc. Figure 4-3 shows this process.

The abundance of food is not the only influence on how a biofilm behaves. Figure 4-4 shows how a biofilm grows, compresses to allow increased flows to pass (thereby bringing more food), and then pulses as the phases repeat. The importance to iron and manganese treatment is the sloughing stage of phase four, referred to as MIR (microbially induced relocation), in which passing water rips pieces from the biofilm. Depending on the extent of biofilm concentrations, slugs of Fe and Mn they carry may hit the treatment process in bunches, perhaps allowing them to pass through the treatment and filtration process.

Figure 4-4 also serves as a reminder of what can happen if Fe and Mn are not removed in the plant, so they can accumulate as biofilms in distribution lines. Such events are often the causes of brown water complaints from consumers. When complaints are received, plant operators naturally consider what changes they can make at their facilities, but short-term measures at the plant may do no good. Brown water complaints can also result from Fe and Mn particulate not removed in the plant that settles and becomes resuspended by changes in flow rate or direction of flow.

The potential for biofilm formation reemphasizes the importance of adequate chlorine residuals throughout the entire distribution system. Consistent levels of 0.3–0.5 mg/L free chlorine improve confidence in the accuracy of chlorine residual measurements and in the generally good condition of the distribution system.

In distribution systems that have not been cleaned for some years while carrying inadequately disinfected waters rich in iron and manganese, flows can be restricted by the formation of microbial corrosion colonies on metal surfaces. These colonies (those mentioned earlier that build covers over themselves to create secure homes) are called *tubercles*. Once tubercles are formed, chlorine cannot penetrate the tough outer shells to kill the corrosive microbes. The most cost-effective remedy is vigorous and repeated swabbing using foam swabs with a coarse outer coating. In some instances, many such swabs may be required to restore clear running water.

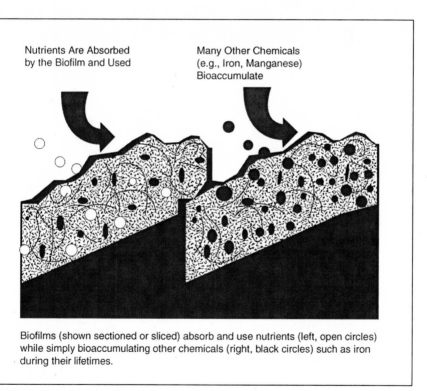

Biofilms (shown sectioned or sliced) absorb and use nutrients (left, open circles) while simply bioaccumulating other chemicals (right, black circles) such as iron during their lifetimes.

Source: Reprinted with permission from D.R. Cullimore. 1993. Practical Manual of Groundwater Microbiology. Chelsea, Mich.: Lewis. Copyright CRC Press, Boca Raton, Florida.

Figure 4-3 How microbes gather iron and manganese

Effect of Microbiological Activity on Wells

When a water well sits idle (i.e., no water is drawn from it) for several days or more, a radical shift in the environmental conditions occurs within any areas fouled by microbiological activity. Elimination of water flows reduces nutrient levels and available oxygen. The biofilm begins to break up, and small colonies of microbes begin looking for a better environment. If pumping from the well resumes, these broken up colonies, now in particulate form, flow easily and create major problems.

Because this series of events continues unseen, the plant operator must rely on tell-tale signs to detect it. Four major indicators (see Figure 4-5) include:

1. Reduced water quality (frequently due to the breakup effects on the biofilm colonies).
2. Increased drawdown due to a "throttling" of the production capacity caused by the gathering of biofilms around the well screen.
3. Increased Fe and Mn concentrations in the biofilms located in the nutrient-rich area created by the startup of the well.

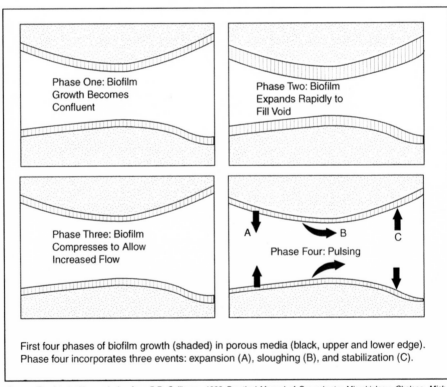

First four phases of biofilm growth (shaded) in porous media (black, upper and lower edge). Phase four incorporates three events: expansion (A), sloughing (B), and stabilization (C).

Source: Reprinted with permission from D.R. Cullimore. 1993. Practical Manual of Groundwater Microbiology. Chelsea, Mich.: Lewis. Copyright CRC Press, Boca Raton, Florida.

Figure 4-4 Brown water complaints: A possible explanation

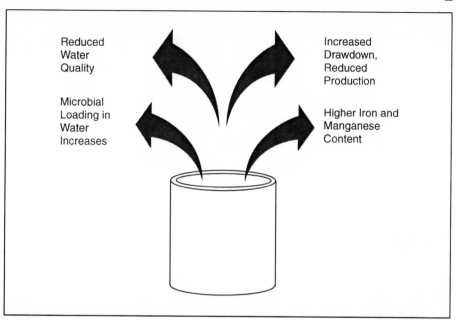

Reduced
Water
Quality

Microbial
Loading in
Water
Increases

Increased
Drawdown,
Reduced
Production

Higher Iron and
Manganese
Content

Source: Reprinted with permission from D.R. Cullimore. 1993. Practical Manual of Groundwater Microbiology. Chelsea, Mich.: Lewis. Copyright CRC Press, Boca Raton, Florida.

Figure 4-5 Some major symptoms of well plugging

   4. Very considerable increases in microbial loadings as the result of disruption
       to biofilms.

Usually, all four of these characteristics occur together, although their order can
vary from well to well.

Many communities draw groundwater from two wells, and some rely on six or
more wells. A community with two wells may alternate well use every month; others
may use certain wells all the time and draw water from reserve wells only in peak
summer periods. To avoid unknown and unwanted additional demands on the
treatment process, some important operating procedures should be followed:

   1. Take weekly samples of the raw water from producing wells at the well sites,
       recording Fe and Mn levels, turbidity, color, odor, and descriptions of any
       visible particulates. The history produced over months of this record keeping
       will highlight any changes.

   2. Never simultaneously start up two (or more) wells that have been sitting idle
       for a week or more. Start one, and take a sample from the well site, then test
       and record the characteristics shown in step 1. Start the next well, test,
       record, and so on. Record the findings in a separate record book or page
       section for each well, so that all information stored in that location applies
       to only one well.

3. After startup, run the well to waste until water quality is very near the level when the well was last shut down, and record how long the flushing took. This period could be minutes, but more likely it will take a much longer time. Obviously, a well with minor biofouling will clean up much faster than one with heavy biofilm contamination.

4. Before combining well waters or switching from one well to another, calculate any needed changes in chemical dosage rates, flow rates, etc. Again, record calculations and changes made; this information will streamline the process each time wells are switched or added.

5. Maintain good record keeping. A well may gradually lose its capacity over a period of years, but the records will show the trend, giving some warning that the well requires rehabilitation or permanent shutdown.

Iron-related bacteria are separated into three major groups. *Gallionella* is termed a *ribbon former*; *Crenothrix*, *Leptothrix*, and *Sphaerotilus* are termed *tube formers*. Group 3 bacteria are termed *consortial heterotrophic incumbents*, meaning that the group includes ribbon formers, tube formers, and any other type of related bacteria that can live in the same environment.

Figure 4-6 illustrates all three major groups. Notice the ribbon former's structure ensures that water spirals down to the base bacteria so that the colony can

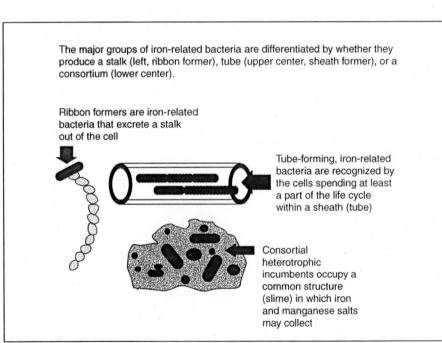

The major groups of iron-related bacteria are differentiated by whether they produce a stalk (left, ribbon former), tube (upper center, sheath former), or a consortium (lower center).

Ribbon formers are iron-related bacteria that excrete a stalk out of the cell

Tube-forming, iron-related bacteria are recognized by the cells spending at least a part of the life cycle within a sheath (tube)

Consortial heterotrophic incumbents occupy a common structure (slime) in which iron and manganese salts may collect

Source: Reprinted with permission from D.R. Cullimore. 1993. Practical Manual of Groundwater Microbiology. Chelsea, Mich.: Lewis. Copyright CRC Press, Boca Raton, Florida.

Figure 4-6 Iron-related bacteria

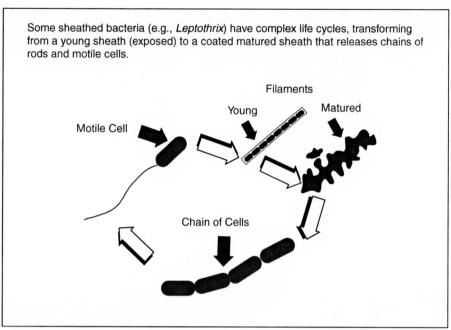

Some sheathed bacteria (e.g., *Leptothrix*) have complex life cycles, transforming from a young sheath (exposed) to a coated matured sheath that releases chains of rods and motile cells.

Source: Reprinted with permission from D.R. Cullimore. 1993. Practical Manual of Groundwater Microbiology. Chelsea, Mich.: Lewis. Copyright CRC Press, Boca Raton, Florida.

Figure 4-7 *Leptothrix* life cycle

extract nutrients. The tube type does the same thing by forcing water through a tube while the colony extracts food.

Figure 4-7 illustrates the growth cycle of *Leptothrix*. (*Motile* means that the cell has the ability to move on its own.)

Sulfur Bacteria, Sulfate-Reducing Bacteria

According to Cullimore (1993, pp. 101–102),

> Sulfur bacteria is an unusual name since it is used to describe two very distinct groups of bacteria. Commonly, the term is used to refer to the sulfate-reducing bacteria (SRB) which is associated with MIC. The term is, however, also used to describe the bacteria in whose activities sulfur plays a major role (e.g., producing elemental sulfur, sulfuric acid). . . . SRB are serious nuisance organisms in water since they can cause severe taste and odor problems and initiate corrosion. These bacteria are called sulfur bacteria because they reduce large quantities of sulfates to generate hydrogen sulfide (H_2S) gas as they grow. . . . The problems generated by H_2S are: (1) it smells like "rotten eggs," (2) it initiates corrosive processes, and (3) the gas can react with dissolved metals such as iron to generate black sulfide deposits.

For the plant operator each of these effects can become a serious nuisance. For example, the sudden appearance of the smell of "rotten eggs" in a water [is one such nuisance]. Usually such events mean that somewhere upstream there has been a major aerobic biofouling which has removed the oxygen out of the water and allowed these bacteria to dominate. Rotten-egg smells are not created by just the SRB group but can also be sometimes generated by other bacteria such as many of the coliforms when suitable conditions occur. Hydrogen sulfide may be found in waters where oxygen is absent and there are sufficient amounts of dissolved organic materials present.

Forcing oxygen (as air) into the water can stress these SRB microorganisms since it is toxic to their activities. Commonly, the SRB protect themselves by cohabiting slimes and tubercles with other bacteria which are slime-forming. These rotten-egg smells will occur more commonly when a water system (or water well) is not used for a period of time. In these cases, the oxygen in the water is used up by various other bacteria cohabiting the biofilm. Once the oxygen is controlled by these other cohabiting members of the consortium, the growth and activities of the SRB group can become rampant. Sometimes this is accompanied by the intense production of the rotten-egg (hydrogen sulfide) smell.

Cullimore's description contains four very important bits of information for the water treatment plant operator.

1. *Hydrogen sulfide may be found in waters with both a lack of oxygen and sufficient amounts of dissolved organic carbons.* The rotten-egg smell in raw well water is a signal of the potential presence of dissolved organic carbons (DOC). When organic carbons exceed 2.0–2.5 mg/L, a percentage of the Fe and Mn present is almost sure to be organically complexed or behave as though it is. When Fe and Mn are complexed with DOC, simple oxidation–filtration may not remove either or both below targeted levels (i.e., 0.3 mg/L iron and 0.05 mg/L manganese). Pilot testing must be done to identify a treatment process that removes iron and manganese to targeted levels.

2. *Hydrogen sulfide is a known reducing agent* (as defined in the glossary). In a manganese greensand filter, for example, hydrogen sulfide can reduce the manganese dioxide coating on the greensand to the manganese form that the treatment process is intended to remove, i.e., Mn^{2+}. This substance slips through the filter into the clear well and distribution system, where it may become a serious nuisance. Again, this problem can occur in a filter at rest (i.e., between periods when water is being filtered), if raw water with sufficient dissolved organic carbons also lacks an oxidant. If this activity continues over long periods, the manganese greensand slowly becomes depleted of its manganese oxide coatings, or the formation of manganese sulfide on the particles inhibits the filter's function. When manganese greensand is stripped of its $MnO_2(s)$ coating, it can perform only mechanical filtering.

3. *Aeration is often the cheapest method to remove H_2S gas from raw water.*
   Hydrogen sulfide can also be neutralized by adding chlorine to the raw water,
   although this method requires high concentrations of chlorine for
   effectiveness. Hydrogen sulfide is also neutralized by hydrogen peroxide,
   although this method is not recommended in Fe and Mn removal processes
   because hydrogen peroxide has an adverse affect on $MnO_2(s)$. Wherever
   possible, H_2S gas should be removed before the addition of treatment
   chemicals.

4. *The presence of SRB is often detected by the formation of visible, shiny grey
   floaters on the surface of the clear well.* The size can vary from 3 mm (0.12
   in.) long to oval-shaped floaters 25 mm (1 in.) long by 13 mm (0.5 in.) wide.
   Once established, these floating colonies of sulfur bacteria can cover 30 to 50
   percent of the clear well's surface. When carefully scooped up in a white
   styrofoam cup, they are seen to carry fine black particulate. Again, once these
   bacteria reach the distribution system, the change in environment, nutrient
   levels, and redox potential can result in taste and odor problems. One section
   of the distribution system may suffer but not another; the problem may seem
   to go away only to show up in another section of the distribution system.
   Treated finished water may be free of any odor at the time it goes into the
   clear well, only to develop tastes and odors while in storage there as a result
   of SRB activity.

BART™: Biological Activity Reaction Tests

One method available to determine the presence of several types of microbiological organisms is the BART™ test.* The test procedure is simple—add a
specified amount of sample water to a BART™ tube, then let it sit and make daily
observations and records of the color, consistency, and location within the tube of
the changes that take place. Comparisons of the observations to the chart included
with each kit allow easy interpretation of the reactions.

BART™ test kits are available for iron-related bacteria (IRB), sulfate-reducing
bacteria (SRB), slime-forming bacteria (SLYM), denitrifying bacteria (DN), nitrifying
bacteria (N), total aerobic bacteria (TAB), and blue–green algae (ALG).

---

*Available from Hach Company, Loveland, Colo.

Chapter 5

Pretreatment

_____ ∎

Common Pretreatment Methods

Aeration

Aeration is often the first pretreatment measure used to prepare water for filtration. Water containing hydrogen sulfide is aerated to remove the gas from the water and release it to the atmosphere. Aeration also helps to oxidize iron. Oxygen, about 20 percent of air, oxidizes iron, though at varying rates. Depending on water pH, temperature, and detention time, and in the absence of organic interference, oxidized iron often forms iron hydroxide, which agglomerates to a relatively large, heavy floc of particles that constitute filterable units. The agglomerated iron is then filtered out in the upper portion of the filter bed made up of coal sized specifically for the site, ranging in effective size (ES) from 0.7 to 1.2 mm (0.03 to 0.04 in.). Experience indicates that coal of this size is usually fine enough to filter out most iron hydroxides and create collisions of oxidized particles that enhance coagulation. At the same time, it is coarse enough to achieve a higher solids loading capacity than the sand or manganese greensand beneath it in most filter beds. If iron oxidized using aeration cannot be captured in a granular media filter, the operator can substitute a coagulant (alum, polymers, etc.) or an adsorption process.

The most common method of aeration is the forced-air cascading tower. Water enters the tower from the top and cascades down over staggered wooden slats while a blower forces air up through the tower from the bottom. The water exits the tower into a detention tank from which it is pumped or allowed to overflow into the filters.

A variation of the cascading tower is cascading steps. In this configuration, water is simply allowed to flow down a channel that resembles a wide set of stairs open to the air. Sufficient turbulence results to dissolve enough oxygen into the water. Another variation is the water spray technique, in which water is sprayed into a detention tank, picking up oxygen from the air in the process.

Common in small treatment plants is the porous-tube type aerator. Compressed air is forced into the center of a porous tube installed in a raw-water line. The air exits the tube in the form of millions of tiny bubbles that are picked up by the water as it flows. This method is effective, but it requires considerable maintenance, since the needle valves controlling the air flow, compressor controls, and bleed lines require frequent servicing. Neglect of needed service often results in sporadic operation, which in turn results in inconsistent oxidation of iron. Failure to determine and follow an appropriate tube-cleaning schedule also results in inconsistent oxidation.

Even smaller plants commonly use a series of venturi devices (sometimes referred to as *hydro-chargers*). The venturi principle takes advantage of suction created by water flowing at high velocity past a small orifice. A simple adjustment screw controls the amount of air drawn into the water flow. These devices are reasonably effective when correctly adjusted, but they require frequent cleaning to remove iron scale and rust buildup.

Chlorination

Chlorine (Cl_2) is usually dosed in one of two ways, as a gas forced into water under pressure or as a solution pumped into the water line by a chemical pump. Chlorine gas is considered to provide 100 percent available chlorine. Calcium hypochlorite is shipped in granular or pellet form and is mixed with water before dosing. The granules or pellets contain 65 percent available Cl_2. Sodium hypochlorite is shipped as a liquid and contains 12 percent available Cl_2; it is dosed either directly from the shipping barrel full strength or after dilution with water in a batch mix tank.

Mixing Cl_2 with water forms hypochlorous acid, $HOCl$. When Cl_2 is mixed with hard water, caustic is also formed, which softens the water and produces precipitates. Many well waters exhibit high hardness and produce troublesome amounts of precipitates when softened. Mixing either hypochlorite product with water should take place in a batch tank, allowing time to settle and stabilize before drawing off the diluted solution into a second chemical container from which doses are withdrawn by a chemical pump. Without the precaution of the two-tank method, the precipitates can be drawn into the chemical pump and cause it to malfunction, or the precipitates can actually be formed inside the chemical pump and foul its operation.

Enough Cl_2 is usually dosed ahead of the filters to provide for oxidation of iron and a free Cl_2 residual of not less than 0.1 mg/L in the water leaving the clear well. The dosage is usually 0.64 mg/L Cl_2 for each milligram per litre of iron in the raw water, plus the free Cl_2 residual, plus any other Cl_2 demands appropriate for the raw water. A Cl_2 dosage substantially higher than the theoretical dosage indicates that the raw water has a hidden high demand for chlorine.

For example, water with a raw-iron content of 5 mg/L has an initial Cl_2 demand of 3.2 mg/L (5 x 0.64 mg/L, the theoretical iron demand), plus possible other minor demands and a free Cl_2 residual of 0.1–0.3 mg/L. The total theoretical dosage is about 3.5 mg/L (using a residual of 0.3 mg/L). If, however, a dosage of 5 mg/L is needed to oxidize the iron (Fe) and provide a free Cl_2 residual of 0.3 mg/L, what accounts for the additional demand?

In surface water, the additional demand often results from the presence of organic materials. In groundwater, the additional demand often reflects the presence of ammonia and hydrogen sulfide (rotten-egg gas); where both of these compounds are found, organic carbons are also usually found. The chemical reaction between iron and Cl_2 is preferential to that between organic carbons and Cl_2; iron and Cl_2 typically (though not always) react to completion in well under 1 min, while organic carbons and Cl_2 typically react to completion in 15 min (although the reaction sometimes takes hours). So, if just the right dosage of Cl_2 needed to oxidize iron is used, none is left to react with organic carbons.

According to Knocke et al. (1990, p. 52), "Free chlorine was much more efficient in its ability to oxidize uncomplexed Fe (II); . . . Although the reaction was not instantaneous . . . efficient Fe(II) oxidation was typically observed within 10 to 15 s (seconds), even under pH 5.0 conditions . . . . A decrease in solution temperature resulted in a noticeable decrease in Fe(II) oxidation rate . . . . However, even under the lowest temperature conditions examined, effective Fe(II) oxidation by free chlorine was observed within 15 s (at pH 6.0) or 90 s (at pH 5.0)." This research was conducted using water temperatures of 2°C (36°F), 10°C (50°F), and 25°C (77°F).

Potassium Permanganate Treatment

Potassium permanganate ($KMnO_4$) is commonly used to oxidize both Fe and Mn. It can also be used to regenerate manganese greensand or pyrolusite filter beds. $KMnO_4$ usually comes as purple crystals in 25-, 50-, and 150-kg (55-, 110-, and 330-lb) drums. The active ingredient is 95 to 99 percent strength, depending on the supplier.

Potassium permanganate reacts vigorously with organic materials, such as powdered or activated carbon, oil, and greases. Carefully store the chemical away from other chemicals to avoid potentially violent reactions. Under some conditions, certain combinations can produce spontaneous explosions. In moist or humid conditions, $KMnO_4$ cakes (i.e., forms solid lumps). After removing sufficient chemical for a new batch, the plastic bag inside the drum should be secured to isolate the chemical from moist treatment plant air. If several drums are purchased at one time, the unopened ones should be stored in dry surroundings on a concrete floor. Potassium permanganate is always dissolved in solution before it is dosed. Typical $KMnO_4$ preparation and dosing systems for large plants are shown in Figure 5-1.

In a large plant, a volumetric feeder continuously adds a measured amount of crystals into a solution tank equipped with a mixer. An eductor draws the prepared solution from the tank and transfers it to the dosing location. Since the flow drawn from the mixing tank by the eductor is usually constant, dosing changes are made by adjusting the feed rate from the volumetric feeder. This adjustment changes dosing by changing the solution strength; the same volume of solution is fed to the treatment process, but the percentage active ingredient strength changes. For continuous dissolving systems like this, the solution strength should not exceed 1 percent.

In a small plant, batch solutions of known strength are mixed as required, usually twice each week or so. The solution is then fed to the raw water using a small chemical metering pump. Changes to the dosing rate (which is the same as the feed

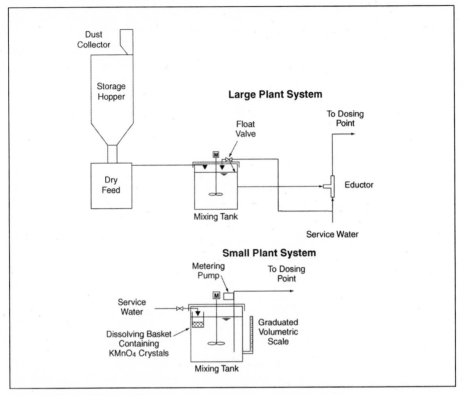

Source: Reid Crowther & Partners Ltd.

Figure 5-1 Potassium permanganate preparation and dosing systems

rate) are made by adjusting the metering pump. Solution strength is usually in the range of 1–3 percent by weight. Solutions are prepared by simply pouring the $KMnO_4$ crystals directly into the batch tank as water is filling it or putting the crystals into a basket or porous sack that is then suspended in the batch tank. Many mixer designs are in use, the most common being a simple blade on the end of a shaft driven by an electric motor.

The rate at which $KMnO_4$ goes into solution is powerfully influenced by water temperature. In room-temperature water (i.e., 20°C [68°F]), the crystals quickly dissolve. A typical small plant, however, uses water drawn from the clear well at a temperature around 5°C (41°F). At this temperature, the slow rate of dissolving requires the operator to leave the mixer running for up to a half hour. The solubility/temperature curve for $KMnO_4$ in Figure 5-2 shows the dramatic effect of temperature on the degree of solubility.

The ideal water for batch mixing is good-quality treated water from the clear well. Under no circumstances should raw water high in Fe and Mn be used for batch mixing, as the Fe and Mn will oxidize and precipitate out of solution. If precipitates

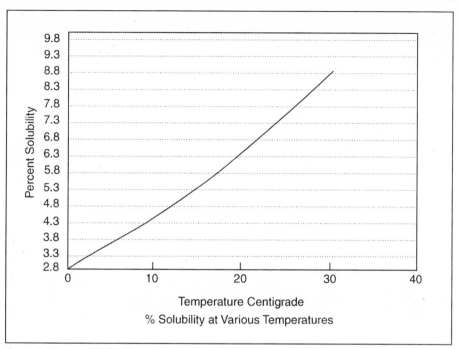

Source: Reid Crowther & Partners Ltd.

Figure 5-2 Percent solubility KMnO$_4$ saturators

build up, they will be drawn into the metering pump and lines, where they may cause blockage or poor pump performance. Residues that collect on the bottom of the batch tank should be removed from time to time.

If space and facilities permit, the best method is the two-tank method. The solution is mixed in one batch tank, allowed to settle, then transferred to a second tank without disturbing accumulated precipitates. The chemical metering pump draws solution for treatment from the second tank.

For a 1 percent solution, 1 kg of KMnO$_4$ is added to 100 L of water. For a 2 percent solution, the ratio is 2 kg to 100 L. Precise oxidation demand rates are given elsewhere in this handbook, but a common rule of thumb calls for 1 mg/L of KMnO$_4$ to oxidize each milligram per litre of iron and 2 mg/L KMnO$_4$ to oxidize each milligram per litre of manganese. Be cautious when applying rules of thumb.

pH Adjustment

Chapter 2 discussed the importance of pH to Fe and Mn removal in a section titled pH Value. Because pH is measured in numbers ranging from 1 to 15, operators can easily fall into the trap of dismissing a difference of 1 pH unit or less as insignificant, especially when they consider smaller differences, such as comparing pH 7.0 to pH 7.3. Remember, however, that pH is measured on a logarithmic scale;

every change of 1 pH unit represents a tenfold difference. For example, pH 8 indicates 10 times more alkalinity than pH 7, pH 9 indicates 100 times more alkalinity than pH 7, and so on. Viewed from this perspective, the importance of pH in a treatment process takes on a whole new meaning.

The rate of Mn oxidation induced by $KMnO_4$ is influenced by pH and temperature. Mn oxidation at pH values between 5.5 and 9.0 generally occurs within 10 s at a water temperature of 25°C (95°F). At 2°C (36°F), oxidation could take 2 min or more, which is considered a long time in small plants whose designs do not incorporate detention time. For many years, the rate of Mn oxidation by $KMnO_4$ was considered especially dependent on pH, so soda ash was added to the raw water to increase its pH from, say, 7.0 to 7.5–7.7. Increasing raw-water pH continues in plants where every second counts.

Aeration of a raw water will likely have no direct oxidizing effect on Mn in solution. By stripping the carbon dioxide (CO_2) out of the water, however, aeration raises the pH, and that change could affect how quickly Fe and Mn can be oxidized.

This handbook cannot address all the potential ramifications of altering pH, but any operator should be aware of any natural changes in raw water pH. Plants should prepare to increase or decrease pH of treated water to improve the overall quality from consumers' points of view.

Other Pretreatment Oxidants: Ozone, Chlorine Dioxide, and Hydrogen Peroxide

Ozone (O_3)

William R. Knocke, of Virginia Polytechnic Institute and State University, conducted valuable research for AWWA and the AWWA Research Foundation to document the effectiveness of ozone (O_3) on oxidation of Fe and Mn. The cost of producing ozone had not decreased enough by the mid-1990s, however, to make this treatment method economically feasible for small plants. The extreme reactivity of ozone also complicates dosage control.

Knocke found that O_3 would oxidize iron almost instantly under typical pH and temperature conditions; however, some types of organic complexing resulted in very poor rates of iron oxidation during O_3 treatment. Reaction times for Mn oxidation were somewhat slower but still in the range of 10 to 30 s. Knocke found that pH played an important role in Mn oxidation, as did temperature. Humic materials (carbon compounds) significantly inhibited Mn oxidation rates. In relatively high doses, however, O_3 does a good job of oxidizing many organic carbon compounds.

This method brings one disadvantage for the water treatment plant operator, in that it does not leave a residual as a continuing disinfectant, as Cl_2 does. However, producers of bottled water for sale in grocery stores often use O_3. They value its strong and rapid disinfecting power, and they need not worry about contamination in a distribution system.

Clearly, extensive knowledge of a raw water's chemical makeup is required in considering O_3 treatment. Even with that knowledge, pilot testing is a must.

Chlorine Dioxide (ClO₂)

Oxidation rates for Fe and Mn are extremely rapid—just a few seconds at room temperatures, and a bit slower at lower temperatures. In treatment with $KMnO_4$ and Cl_2, however, the presence of dissolved organic matter (i.e., organically complexed Fe and Mn), tends to slow oxidation rates. Such a situation affects the use of chlorine dioxide (ClO_2) as an oxidant, since it differs from the other common oxidants in two ways.

First, ClO_2 reacts very rapidly with many organics, which makes the organic material an oxidant competitor to Mn. Extensive testing is required to determine the amounts and types of organic materials present in a raw water before dosage rates can be established. Also, a treatment plant using ClO_2 would need some means of tracking increases and decreases in organic levels, not an easy activity for a small plant.

Second, Knocke found that the interaction between ClO_2 and reduced Mn yields chlorite. Concerns regarding the potential health implications of residual chlorite concentrations might limit the usefulness of ClO_2 as a Mn(II) oxidant.

Chlorine dioxide is always used in a gaseous form. The product is generated at the point of use.

Hydrogen Peroxide (H₂O₂)

Hydrogen peroxide (H_2O_2) is the weakest of the five oxidants discussed in this chapter. It generally yields unsatisfactory reactions with organically complexed Fe, and it has no practical oxidative effect on Mn(II).

Testing Equipment

Treatment plant operators test (or analyze) water for several reasons. They must identify concentrations of Fe and Mn, as well as potential interference with treatment, to choose the most appropriate treatment methods. Levels of Fe and Mn must be determined so that chemical dosages can be calculated. During treatment, spot testing confirms the appropriateness of earlier dosage calculations or indicates a need for increased or decreased dosage. Even following precise dosing, spot testing can alert an operator to changes in raw-water quality and any need for further adjustments to ensure that consumers receive water that meets or exceeds quality objectives.

Although regulatory agencies may require certain testing activities, this section deals mainly with tests chosen by water treatment plants to assure attainment of water quality objectives. Tests for elements and compounds related to human health and toxicity and tests for general chemical parameters are performed in professional laboratories using sophisticated equipment and techniques; these advanced functions are well beyond the capabilities of most portable and small system labs.

Particle Counting Equipment

In the mid-1990s, particle counting began coming into its own. Although the technology remains very expensive, particle counters have already dropped substantially in cost, and they may become cost-effective tools in many plants within several years.

However, debate continues among professionals about whether or not particle counting is an appropriate tool for monitoring and measuring Fe and Mn removal processes. Some theories seem to indicate that particle counters generate different information than more traditional methods provide, although new data do not necessarily offer an improvement over good knowledge of soluble and particulate forms of Fe and Mn in raw and finished water correlated to turbidity. Today's costs for residual testing and turbidity counts remain below the capital, operating, and maintenance costs of a particle counter.

Although some researchers have reported findings based on data from particle counters in *Journal AWWA*, only limited documentation details the role of particle counters in Fe and Mn removal processes. To evaluate this technology further, read appendix B, and then draw a personal conclusion on its usefulness after conducting a cost/benefit analysis.

Test Kits

Some communities still measure residuals using a color comparison method. This method requires the operator to add a reagent to a water sample, wait for a color change to develop fully, and then visually compare the result to a standard solution or some device that displays many shades of color. Many sources of inaccuracy can cause problems with this method. The comparator disk, wheel, or multidisks may be dirty, or the reading may be taken in bright light one day and lower light the next. The judgment of one employee may not match that of another.

A step up from the color comparator is the colorimeter. This device measures a sample's color intensity on a scale calibrated using standards of known concentrations. The range normally includes a blank (concentration = 0) made from distilled water to adjust the instrument's zero point. The standard or sample is placed in a test tube or cuvette (a test tube with a square or rectangular section) which fits inside the colorimeter well. A light beam passes through the test tube or cuvette, and a photocell measures the amount of light reaching the other side. As everything else is made equal, color intensity is proportional to the amount of light absorbed by the liquid inside the test tube or cuvette.

Maximum absorbances, and therefore sensitivity, can be obtained by using a light beam of a color complementary to that of the developed sample. For example, if the developed sample coloration is purple, its concentration would be measured using yellow light. Some colorimeter models come equipped with separate colored glass filters that can be inserted between the light bulb and the well. Other models mount these filters internally on a drum. Rotating the drum places different filters in the light path.

Both of these methods employ old technology. The updated technology of the spectrophotometer offers many advantages. The spectrophotometers of the mid-1990s employ diffraction prisms and offer preprogrammed capabilities to perform many dozens of different tests, display digital readouts, and carry out print and computing functions.

Methods of Testing for Soluble Metals

Testing for Iron

Ferrous iron (Fe^{2+}) can be measured based on color changes developed by a number of reagents. One such reagent is 1,10-phenantroline. When Fe^{2+} comes into contact with phenantroline, a reddish-orange coloration develops with an intensity directly proportional to iron concentration, assuming an excess of phenantroline remains. Colorimeters use a bluish-green filter (510 nm light wave length).

Ferric iron (Fe^{3+}) is reduced to Fe^{2+} for measurement, which is done using another reagent, hydroxylamine. Because the reaction and color development must take place at a pH of about 3, a buffer solution of ammonium acetate is used. This reaction takes about 15 min. Other recognized methods are based on the reaction of iron with ferrozine and TPTZ.

Several substances may interfere with the determination of Fe concentration. The list includes strong oxidizing agents (such as chlorine and permanganate), cyanide, nitrite, phosphates, chromium, zinc (at concentrations more than 10 times the iron), cobalt and copper (at concentrations of more than 5 mg/L), nickel (at more than 2 mg/L), bismuth, cadmium, mercury, molybdate, and silver. Natural coloration due to organic matter may also interfere with this measurement, an effect that is difficult to identify. Some reagents contain additional chemicals to prevent certain interference. An operator should compare raw-water analyses conducted by a professional laboratory to the test manufacturer's list of interference problems to assure accurate testing. Choose the kit with the reagent that offers the greatest chance of providing an accurate determination.

This choice weighs options from a number of test kit and reagent suppliers. The test kit and method selected must allow determination of residuals in the concentration range required. For example, some kits test Fe levels only from 0 to 0.1 mg/L, while others may test from 0 to 10 mg/L. Generally, the most precise determination comes from a test that applies to a narrow range. For example, if the raw water's Fe concentration will be near 3.0 mg/L, a kit designed to test in the 0 to 4.0 mg/L range is a better choice than one designed for 0 to 50 mg/L.

Testing for Manganese

One method for measuring Mn oxidizes the substance to the soluble permanganate form. In this test, a colorimeter uses a yellowish-green filter (525 nm light wave length). Suspended and colloidal manganese must be dissolved using acid. Some interferences can be avoided by adding mercuric sulfate and silver nitrate. The standard oxidant in this method, sodium persulfate, is used under boiling conditions; oxidants such as periodate do not require boiling.

Several substances can interfere with testing to determine Mn levels. Some are organic matter, turbidity, and other sources of color. As with Fe test kits, some reagents contain chemicals that compensate for interferences.

The Pan Method test kit is popular with small-plant operators because addition of three reagents provides a reliable color indicator of the Mn level. Again, it is important to select the appropriate concentration range; kit capabilities range from 0–0.6 mg/L to as wide as 0–10.0 mg/L.

If using the Pan Method kit, allow 10 min for full color development if Fe in the sample exceeds 5 mg/L. If sample hardness as calcium carbonate ($CaCO_3$) exceeds 300 mg/L, add rochelle salt solution as recommended by the kit manufacturer (Hach Company).

Testing Oxidation Efficiency

Removal of Fe and Mn from drinking water often entails precipitation by oxidation then separation of the solids by granular media filtration. If water quality objectives for Fe and Mn are not met, the level of oxidation is an important variable to check. If a high level of oxidation is taking place, investigation must determine why the precipitate is passing through the filter.

Iron may precipitate as iron hydroxide and Mn as manganic dioxide. Both may appear in colloidal form (defined in the glossary). The concentration of oxidized metals can be determined by filtering a sample through a membrane of known pore size before performing one of the Fe or Mn tests described earlier.

Apparatus for membrane filtration (sometimes called *millepore testing*) is shown in Figure 5-3. A membrane of known pore size is placed in the holder, and the sample is poured into the container above it. A hand vacuum pump removes enough air from the filtration flask to draw the sample water through the membrane, typically at less than 35 kPa (5 psi). Membrane filter disks (most of them round in shape) come in pore sizes as small as 0.22 μm. (A micron is 1 one-thousandth of a millimetre.) A common size is 0.45 μm.

If a sample drawn through a 0.45 μm filter still has a high percentage of the raw-water levels of Fe and Mn, a granular filter certainly will not remove them. Such a sample would have been taken after dosing with an oxidant (O_2 [aq], Cl_2, or $KMnO_4$) leaving the question, Is the particulate very fine or colloidal in size, or were the Fe and Mn not oxidized? Sometimes both O_2 (aq) and Cl_2 can take from several minutes

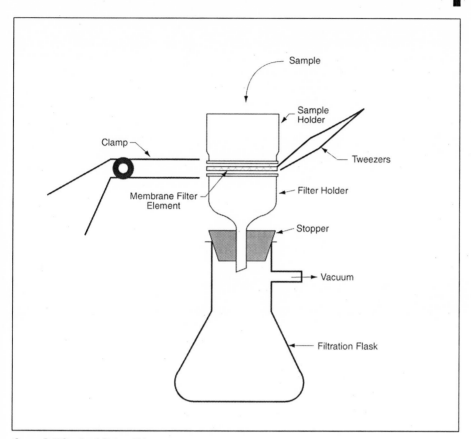

Source: Reid Crowther & Partners Ltd.

Figure 5-3 Dissolved metal analysis membrane filtration

to hours to complete Fe and Mn oxidation; sometimes both Cl_2 and $KMnO_4$ can oxidize Fe and Mn quickly, but the resulting precipitate may remain very fine or colloidal in size.

Further analysis would take a sample of oxidized raw water from the plant and add measured amounts of a coagulant, say alum or a cationic polymer, to a series of jars. Stir gently for several minutes (say 15 min), allow the samples to settle for several minutes (say 15 min more), then filter through a 0.45 μm membrane again. If the filtrate then tests very low in Fe and Mn, the particles in the first sample were very fine or colloidal in size. This result indicates that a coagulant (selected on the basis of further pilot testing) should be a part of the pretreatment process. In some instances, oxidized Fe and Mn can even be subcolloidal in size and extremely difficult to flocculate, even using coagulants.

If the addition of a coagulant makes little or no difference, the Fe and Mn likely remain in the sample because they were not oxidized. In this case, a series of experiments should test each of the three potential oxidants (i.e., O_2 [aq], Cl_2, and $KMnO_4$), allowing a reasonable amount of time for the oxidation reaction to go to completion. These experiments should determine how long complete oxidation takes. The challenge then is to determine if that amount of time can be made available within the treatment sequence.

Most small plants can allow only very short detention times. Some plants have room to install tank capacity that can lengthen detention time by several minutes or up to a half hour. If detention time experiments indicate a need for hours of detention time, some other treatment process will likely need to be considered.

Membrane filter tests using a number of different pore sizes can help to establish a particulate size distribution. For example, if the finest test membrane is 0.45 μm and the coarsest is 10.0 μm with two or three sizes in between, testing a sample of oxidized water may show a percentage of particulate removed by each pore size. These results are compared to the smallest particulate size that can be removed in the plant's filters.

Testing for Chemical Oxidant Demand

Measurements of chemical oxidant demand give important guidance to determine the dosage of chemicals required for proper treatment. The three most commonly used oxidants are O_2 (aq), Cl_2, and $KMnO_4$. The demand test generally requires addition of the oxidant in known quantities, a reaction time similar to what occurs in the WTP, and determination of the oxidant residual after completion of the reaction time. The oxidant demand is the difference between the dose added and the residual.

O_2 (aq) demand. Because a residual of oxygen in water does not have an important effect on health, little effort is usually taken to determine just how much is required. Frequently a dissolved oxygen saturation level of about 60 percent is sufficient for iron oxidation. An operator may want to know, however, just how much dissolved oxygen results from any aeration process. Too much could cause problems such as accumulation in an underdrain, and resulting media disruption during backwash. Under certain circumstances, air could be forced out of solution in the filter media and reduce filter run lengths. A surplus of oxygen might also stimulate unwanted microbiological activity. In particular, an occasional measurement of dissolved oxygen is wise during compliance inspections, with adjustments to aerators as necessary.

Cl_2 demand example. For this example, suppose that four 1,000 mL samples are placed in four 1 L beakers. The first receives 1 mL of calcium hypochlorite solution containing 1 g/L of Cl_2. The second beaker receives 2 mL of the same solution, the third 3 mL, and the fourth 4 mL. The beakers are then kept in the dark to prevent photochemical reactions. If the plant's hydraulic detention time between chemical dosing and filtration is about 10 min, the samples are tested for total chlorine residual after that time. No chlorine residual is detected in the first beaker. The chlorine residual in the second beaker is 0.05 mg/L. The residual in the third beaker is 1.1 mg/L. The fourth sample is not analyzed.

To find the raw water's chlorine demand, first determine how much chlorine was added to each beaker. Since the dosage contained 1 g/L of Cl_2, the chemical equals 1,000 mg/L, or 1,000 mg/1,000 mL. The first beaker of sample received 1 mL of this solution in 1,000 mL (1 L). Therefore the dosage of chlorine for the first beaker is:

amount of Cl_2 added = amount of solution added x solution concentration
= 1 mL of solution x 1,000 mg Cl_2 per 1,000 mg of solution
= 1 mg Cl_2
chlorine dosage = amount of Cl_2 added/sample volume
= 1 mg Cl_2/1,000 mL of sample
= 1 mg/L

The dosage for the other beakers can be calculated similarly as 2, 3, and 4 mg/L, respectively. The zero residual concentration of chlorine in the first beaker indicates that the chlorine demand in the raw water was at least 1 mg/L, but probably greater. The demand for the second beaker can be calculated as follows:

chlorine demand = chlorine dosage – chlorine residual
= 2 mg/L – 0.05 mg/L
= 1.95 mg/L

The chlorine demand for the third beaker can be calculated similarly as 1.9 mg/L. The similarity of the chlorine demand measured in the second and third beakers confirms that the raw water has a Cl_2 demand of 1.9 mg/L. The average of the two could have been taken, but they are obviously so close that this step is not needed. The sample with the greatest residual concentration (i.e., 3 mg/L) is likely the one with the least error. As the second and third beakers indicated the same chlorine demand, the test need not continue to analyze the fourth sample. The chlorine demand in the raw water is established as 1.9 mg/L.

Reagent solution preparation. To prepare a Cl_2 reagent solution with a concentration of 1 g/L, 1.55 g of calcium hypochlorite is dissolved in 1,000 mL of distilled water contained in a clean, dark bottle. This solution should be prepared immediately before use in the test, and any remainder should be discarded after all tests have been completed for the day. Calcium hypochlorite is unstable and cannot be stored for a long time. (Sodium hypochlorite could have been used, but it is an even more unstable compound.)

$KMnO_4$ demand example. To determine potassium permanganate demand, proceed similarly to the chlorine demand test as indicated in the previous example. However, instead of testing for residual $KMnO_4$, coloration is observed after the reaction time.

Oxidized Fe and Mn form a reddish-orange to brown precipitate. Residual $KMnO_4$ gives the solution a pink coloration. Therefore, the sample demand falls between the dosages corresponding to the two consecutive beaker samples for which one has no pink coloration and the next one does have a pink coloration. To fine-tune the demand determination, repeat the test using dosages intermediate between the two.

For example, if the first test found a $KMnO_4$ demand between 1 and 2 mg/L, the next test could use dosages of 1.2, 1.4, 1.6, and 1.8 mg/L. If the one dosed with 1.2 mg/L $KMnO_4$ showed no pink, but the one dosed with 1.4 mg/L did show pink, the demand could be reported as 1.3 mg/L. On the other hand, if all beakers showed a pink coloration, the demand would be 1.1 mg/L. (Remember that the beaker with 1 mg/L was not pink in the first test, so the demand must be greater than that amount.)

To prepare a 1 g/L reagent solution of $KMnO_4$, dissolve 1.05 g of $KMnO_4$ in 1,000 mL of distilled water in a clean, dark bottle. This bottle should have a plastic cap, a label indicating its contents, and the date of preparation. This solution can be used for about a month if kept in a cool, dark place.

Chapter 6

Filtration

∎

This handbook details a process for Fe and Mn removal composed of three equal parts: pretreatment of the raw water to produce a filterable precipitate, filtration, and sustainability of the filtration function. Decades of troubleshooting Fe and Mn removal plants has demonstrated time and again a major adverse impact by fouled media beds resulting from poor underdrain design. The section on underdrains gives sufficient information to broaden the scope of troubleshooting.

Common Filter Types

The two most common types of filters in use today are open-top, high-rate gravity filters and enclosed steel pressure filters. Gravity filters are of two types, high-rate and slow sand filters. Strong demand for water leads plants to avoid slow sand filters. Most pressure filters are steel, although those in very small plants are sometimes constructed of fiberglass. Any material that has a long life, can withstand the pressure, and does not give any taste, odor, or contamination to the water would suit the needs of pressure filter construction.

The same filter media can be used in either of the two common filter types. All filters require some type of underdrain to collect filtered water and distribute backwash water, and air scour capabilities can be provided in both types. Historically, gravity filters have used the block type of underdrain and less often porous plate, strainer/false bottom, and pipe header/lateral designs, along with variations and combinations. Pressure filters have used false bottoms with strainers or distributors or hub/lateral configurations.

From an operator's perspective, the open-top, gravity filter is preferable to the enclosed pressure type. This design allows the operator to see backwash flows, formation of mud balls, slow formation of dead spots, and fluidization of the media, which improves operating practices and corrective maintenance. The rule "out of sight, out of mind" too often applies to pressure filters, and operators do not see developing conditions until serious problems emerge.

Underdrains

Early forms of underdrains were simply hollow, square-ended ceramic blocks laid end to end and cemented in place. (See Figure 6-8 later in the chapter.) The top side had a pattern of holes up to $1/2$ in. in size. In order to prevent loss of sand into the underdrain, clear well, or distribution system, several layers of graded gravel topped the blocks, starting with a $1^1/_2 \times {}^3/_4$ in. size, decreasing in each layer to a final top layer of $^1/_4 \times {}^1/_8$ in. size (sometimes smaller) called *pea gravel*. The gravel layer, from 1 ft to 18 in. in depth, served two purposes. First, it prevented loss of sand into the underdrain and beyond. Second, it acted as a distributor of backwash water. In theory, water forced up through such a bed would move laterally (horizontally) about 1 ft for each foot in vertical rise. This design was an early attempt at reducing maldistribution of backwash water.

Even early filter underdrain designers recognized the need to distribute backwash water evenly to maintain filter run lengths and water quality and to get the maximum life out of a filter bed. Many attempts to improve this design gained some ground; however, even relatively minor maldistribution of backwash water clearly had a negative effect on media beds over time.

The single most important contributor to media bed failure and deterioration in water quality is accumulation of particulate in the filter resulting from inability of the underdrain to deliver even backwash flows. Secondary contributors are inadequate backwash flow rates and a lack of air scour to loosen particulate before it is washed away to waste. As increasingly efficient chemicals are developed to agglomerate particulate to filterable size, growing attention has focused on even backwash distribution, adequate backwash rates, and air scour capability.

For this reason, the following section discusses the most common types of underdrains and variations. The ability to keep a media bed clean is of paramount importance to sustained operation of any Fe and Mn removal process.

Types of Filter Underdrains

The many types of granular media filters used in municipal and industrial water treatment can be grouped into two distinct categories: gravity and pressure filters. A gravity filter is an atmospheric tank (either concrete or steel) that relies on the difference in elevations between the inlet and outlet to provide driving force that pushes water through the granular media. Gravity filters can be either round or rectangular tanks.

Pressure filters are totally enclosed vessels (manufactured from steel or fiberglass) operating at pressures potentially substantially higher than atmospheric pressure. The actual pressure differential across such a filter bed is identical to that of a gravity filter for the same applications.

Some commonly seen types of granular media filters are:
- Open-top gravity type (Figure 6-1)
- Low-pressure self-backwash storage gravity type (Figure 6-2)
- Vertical pressure type (Figure 6-3)
- Horizontal pressure type (Figure 6-4)

Every filter must include is an underdrain system, though. The underdrain sits just below the filter media and preforms three functions:

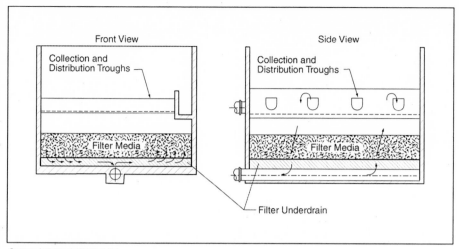

Source: Anthratech Western Inc.

Figure 6-1 Open-top gravity filter

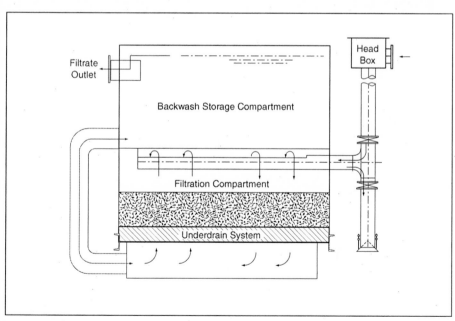

Source: Anthratech Western Inc.

Figure 6-2 Self-backwash storage filter

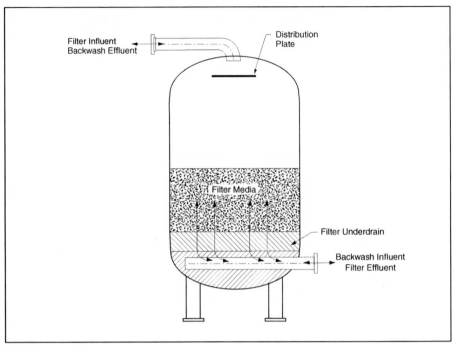

Source: Anthratech Western Inc.

Figure 6-3 Vertical pressure filter

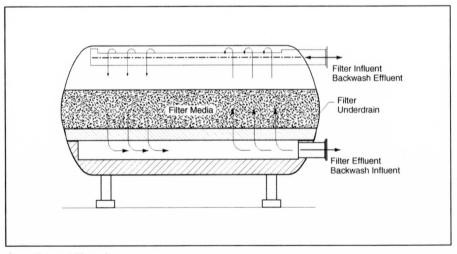

Source: Anthratech Western Inc.

Figure 6-4 Horizontal pressure filter

- Collect filtered water evenly (an easy task)
- Prevent passage and loss of fine media (a more difficult task)
- Distribute backwash flow evenly (a very difficult task)

Underdrains more and more frequently must perform some additional functions, as well. For example, an inherent air scour capability is increasingly recognized as a desirable feature. Another frequent requirement is the capability to use high-rate media configurations (filter coal/sand/garnet or ilmenite) without disruption to this layering. Another strong desire in the industry is an underdrain that does not require gravel layering, but does not eliminate gravel, which would open the door to other problems like strainer failure or plugging. An underdrain without gravel and without strainers would really be an ideal design.

Dozens of different types of underdrains are used in the industry. Many designs are proprietary, but many are not. Current designs can be categorized into five main types:

1. Hub lateral (with variations)
2. True header lateral (with variations)
3. Tile bottom (with variations)
4. False bottom or plenum type (with variations)
5. Newtech variable-orifice type with air scour

Category 1: Hub Lateral Underdrains (with Variations)

This type of underdrain is commonly used in small, vertical pressure filters and greensand filters up to approximately 1.8 m to 2.5 m (6 ft to 8 ft) in diameter. It is the least expensive type of underdrain and generally the least effective performer. In this configuration (Figure 6-5), laterals radiate out like spokes on a wheel from a central hub. As the laterals move farther apart along their lengths, uneven backwash distribution becomes a certainty.

No header maldistribution problem develops as the hub takes the place of a header, and symmetrical laterals evenly accept backwash. To spread this backwash through the filter bed, however, the laterals require control of distribution, increasing water delivered farther from the center, as each lateral must backwash an increasing bed area. This adjustment is sometimes made by varying orifice spacing or slot sizes along each lateral's length.

A variation (Figure 6-6) is a double hub lateral. This design cuts in half the area each lateral is required to cover, but it provides only a partial solution.

A hub lateral underdrain can compensate to a degree for differential upward backwash velocities by incorporating gravel over the underdrain, but this solution is limited in gravel depth by the side wall height restrictions. Sometimes laterals are designed with fine slots or with strainers to eliminate gravel or minimize the depth of barrier gravel. This arrangement ignores the importance of gravel as a diffusion layer, and loss of that diffusion results in uneven backwash velocities through the media.

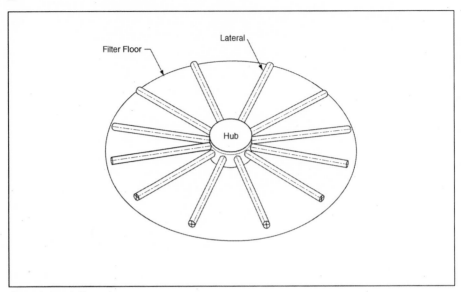

Source: Anthratech Western Inc.

Figure 6-5 Hub lateral

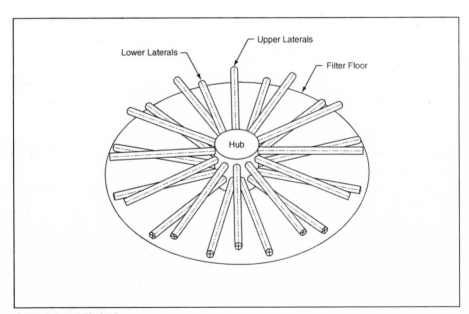

Source: Anthratech Western Inc.

Figure 6-6 Double hub lateral

To minimize cost, the laterals are often made of polyvinyl chloride (PVC) pipe. While this material is inert to corrosion, it may break at connections after prolonged exposure to cold water and flexing during backwash cycles. Breaking of PVC laterals presents a serious problem when it occurs.

Category 2: True Header Lateral Underdrains (with Variations)

These types of underdrains are reasonably common in large, vertical pressure filters, horizontal pressure filters, and some open-top gravity filters. They are frequently used in "packaged" water plants.

In its simplest form (Figure 6-7) this design consists of a pipe header running down the center line of the filter with pipe laterals connected from each side of the header. Usually the pipe laterals have orifice holes and graded gravel layers block fine media passing through. This arrangement achieves a degree of backwash flow distribution.

Individual filters display many variations on this theme. Sometimes, a header runs down the side of the filter with laterals connected to one side. The header may be external to the filter, perhaps a concrete conduit with laterals running through the wall and across the filter. Sometimes the laterals are slotted or wrapped with stainless steel mesh. Some are equipped with strainers, wedge wire or plastic wedge, or plastic cloth wraps. Some designs have air scour capabilities, usually using strainers with air metering tube extensions.

A common traditional variation, the Wagner underdrain, employs a down-the-center box conduit, usually of transite, with transite pipe laterals. Between the lat-

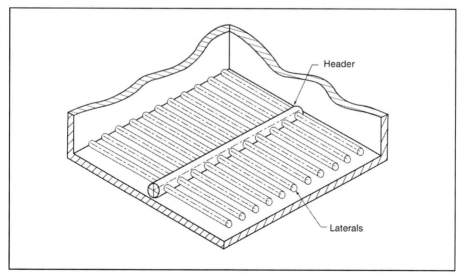

Source: Anthratech Western Inc.

Figure 6-7 True header lateral—centered header

erals, precast blocks are shaped to deflect backwash flow from the laterals upward through slots in the concrete blocks. The installation is gravel layered, and the flat top of the center conduit is usually fitted with strainers or nozzles to allow water to pass rather than allowing this area to remain as dead space.

As mentioned, some designs have air scour capabilities, but many older versions do not. To add air scour to an older system with gravel layering, a separate air header with laterals usually would lie near the bottom of the silica sand media. This placement is necessary, because blowing air through the gravel could disrupt the integrity of gravel layers.

These types of underdrains have been used for many years. While some operate reasonably free from trouble, many are prone to uneven backwash distribution in both header and laterals. Air distribution problems and gravel disruption may reduce effectiveness, as could a sudden release of air trapped in the underdrain system. This last problem can occur even in filters without air scour capabilities.

Category 3: Tile-Bottom Underdrains (with Variations)

The most common designs of this type employ Leopold fired clay, but lately Roberts and others have offered nearly identical products. Gravel layering is required (Figure 6-8). Generally, surface and/or subsurface washers are used, since the blocks cannot accommodate air scour.

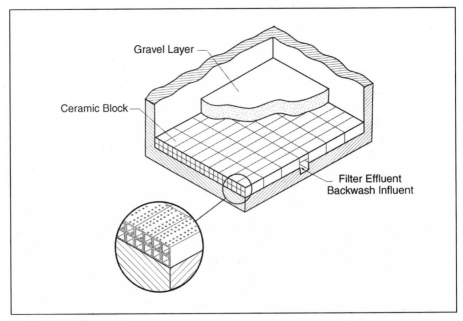

Source: Anthratech Western Inc.

Figure 6-8 Ceramic block underdrain

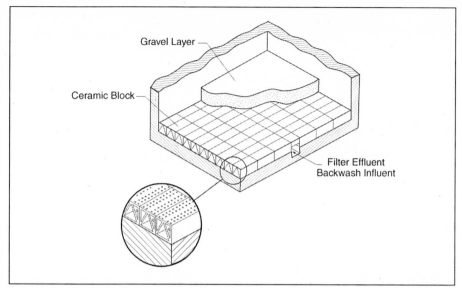

Source: *Anthratech Western Inc.*

Figure 6-9 Universal block underdrain

Variations employ concrete block and Leopold "Universal" plastic block (Figure 6-9), which has air scour capability. An old design, the Miller block, also appears.

Blocks placed in rows form channels that function as laterals. They are left open to a sunken concrete flume running across the front end or down the center line, which serves as the filtered water collector and backwash water header.

The Leopold block, a very old design dating to the 1930s, has returned to popularity. For a period of time (roughly 1950–1965), it lost out heavily to false bottom designs with strainers, which were intended to eliminate the need for gravel. Problems with strainer failure have resulted in a swing back to the Leopold block. In the past few years, the Leopold Universal plastic block has been promoted instead of the clay block, since it offers air scour capability.*

These types of underdrain systems are generally limited to rectangular, concrete open-top gravity filters, and they are not applicable in pressure or low-pressure, self-

---

*The Leopold block prevents gravel layer disruption from air scour through an hourglass gravel configuration. The gravel varies from coarse to fine and back to coarse, a very tedious system to install that has been only moderately successful using low volumes of air. A recent variation is a universal block installed without gravel using an IMS cap of porous plastic beads sintered together. A few initial installations had some failures when the method of attaching the caps to the blocks failed, but subsequent installations have achieved better results. However, some new data show that filter media finer than 0.8 mm effective size pass the cap, eliminating the use of a polishing layer of filter media.

backwashing storage types. Many tile bottom underdrains run trouble-free year after year. In others, though, air upsets the gravel layering, allowing fine filter media to intrude into the block rows. This process is a common form of failure for these types of systems.

Tile block underdrain designs include no header corrections to ensure even distribution into the block row laterals, allowing very significant header-to-lateral maldistribution. Also, installation of such an underdrain is typically a tedious and costly process.

Category 4: False Bottom (or Plenum) Type (with Variations)

A great variety of designs fit in this category. Some require gravel, some do not. Some have strainers, some do not. Some can air scour, and some cannot. Space does not allow detailed descriptions of all these variants, but a list shows the diveristy:

- Wheeler bottom, made up of concrete block or cast in place (This design uses ceramic balls in inverted pyramid depressions; see Figure 6-10.)
- Other cast-in-place variations: Paterson-Candy, Eimco, Ecodyne, and others (Figure 6-11)
- Precast block on pedestals: Infilco Fre-flow, Eimco, Ecodyne, and others
- Asbestos cement or composite block on legs: Infilco Fre-flow and others
- "Tee-Pee" design, consisting of precast concrete in an inverted V shape
- Porous plate design: Carborundum and others
- Double dish (This version is used sometimes in vertical pressure filters.)
- Painted steel or galvanized false bottom plates with strainers from Graver and various suppliers (All self-backwashing storage low-pressure filters have this type of underdrain; see Figure 6-12.)

Some of these underdrains offer much better performance than others, but experience has revealed very frequent problems. Some typical difficulties are:

- Gravel upsets allow fine media to pass.
- Structural failures may include corrosion/uplifting of blocks and inversion of Wheeler balls.
- Strainers may fail to screen out filter media.
- Strainers may plug.
- Uneven backwash distribution may lead to mud balling, channeling, and short-circuiting.
- Poor air scour distribution may reduce effectiveness.
- Explosive air release may result from inability to cope with backwash air entrainment or air release due to negative head operation.
- Media upsets are common, and some filters cannot use high-rate media.
- Due to jetting and roiling action, strainers often require gravel to provide distribution of backwash flow, although they were developed to eliminate gravel.

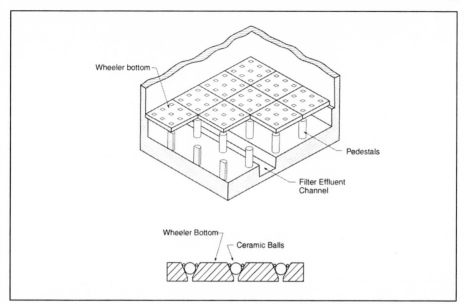

Source: Anthratech Western Inc.

Figure 6-10 Wheeler bottom

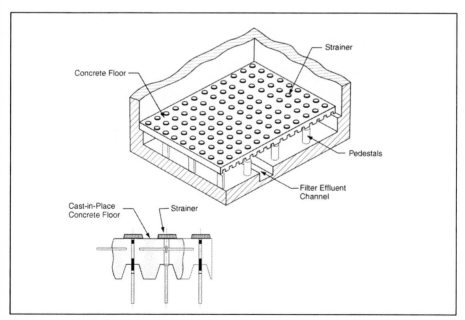

Source: Anthratech Western Inc.

Figure 6-11 Cast-in-place bottom with strainers

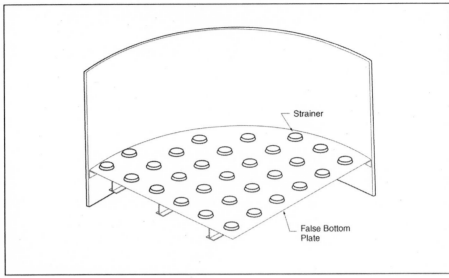

Source: Anthratech Western Inc.

Figure 6-12 False bottom plate with strainers

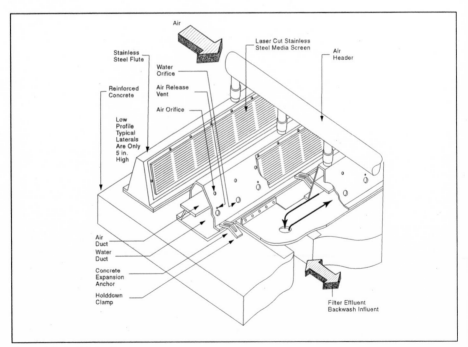

Source: Anthratech Western Inc.

Figure 6-13 Newtech variable-orifice type with air scour

Category 5: Newtech Variable-Orifice Type with Air Scour

J.B. Hambley set out to develop a filter underdrain that would overcome the multitude of problems with current underdrains. The resulting Newtech underdrain system basically combines two header lateral systems (one for air and one for water) into one compact lateral. Each lateral is divided into an upper chamber to carry scouring air and a lower chamber to collect filtrate and distribute backwash water. The locations of the air and water headers are flexible within the filter.

The underdrain system (Figure 6-13) is typically fabricated from stainless steel to take advantage of its structural strength and resistance to corrosion. A reasonably flat floor is required for installation of the underdrain system, which is a variant of the header lateral design. No plenum or false bottom is required.

In a large filter, the header typically is a flume or conduit formed in the concrete either crossing the inside front of the filter cell or running down the center along the length of the filter. It can also be an external conduit feeding the end of each lateral. The laterals are held to the filter floor with anchor bolts set in concrete adhesive and hold-down clamps.

In a small filter, the header is usually a light-gauge stainless steel pipe with stub risers and a sealing plate. These headers are usually encased in concrete to form the filter floor. Again the laterals are held to the filter floor by anchor bolts set in concrete adhesive or J bolts installed in the sealing plate. Each header design uses variable orifices to distribute flow proportionally to each flute lateral.

North American practice for air scouring (where warranted) has called for filtering at rates of 10.2–20.3 (L/s)/m^2 (2–4 SCFM/ft^2) for 3–5 min prior to hydraulic backwash. This procedure has typically applied to dual-media filters involved in depth filtration for which surface washers cannot reach deeply embedded solids. Subsurface washers just above the filter coal/sand zone also affect only a fairly limited zone, and they are troublesome to maintain. Hydraulic backwash alone is quite a gentle action; the much more violent action of air often helps to keep the entire filter bed clean by loosening adherent sticky sludges containing algae, clay, manganese, or lime softening precipitates.

The Newtech underdrain is readily adapted to air scour in a rather unique way. The flute lateral is fabricated with a division plate between the upper air channel and the lower water channel. This is a very important innovation. Air is introduced into the upper "conduit" from either above or below. Air introduction quickly displaces all the water out of the air conduit section, allowing the air to exit.

Another mode of air scour pioneered in Europe began to gain popularity in the mid-1990s, appearing, for example, in the Los Angeles Aqueduct Plant and the Salt Lake City water plant expansion. In this mode, air scouring occurs simultaneously with backwash water addition. Air introduction produces media bed expansion of only 5–8 percent, with just enough backwash flow to provide the sweep velocity needed to remove accumulated solids.

A hazard of simultaneous air/water backwash is loss of media carried to the waste sump by combined flows. Trough guards minimize media losses by separating air from the media before the backwash effluent enters the waste line. More commonly, the filter level is reduced to just above the media bed, the simultaneous mode is turned on, and then air is turned off just prior to the backwash flow reaching the lip of the backwash trough.

Backwash requirements are typically half the level of conventional filters. Special consideration is given to media type and sizing to avoid floating off filter coal. Often such systems employ single media (i.e., sand only or filter coal only) to avoid a final high-rate backwash to restratify dual media.

The Newtech underdrain also avoids another, often unrecognized, contributor to backwash maldistribution common in a number of filter underdrain types: trapped pockets of air. This problem can lead to variability in orifice cross-flow velocities, which often causes even identical filter cells to exhibit different backwash characteristics.

The laterals in the Newtech underdrain system have small air vent holes strategically placed to prevent air from being trapped. Most underdrain systems have no way of coping with air entrainment or explosive air releases. Trapped air can seriously affect backwash distribution, and explosive air release is violently destructive to gravel layering, producing high float-off losses of filter coal.

The standard shape of flute lateral coordinates with gravel layering, sizing the initial gravel layer to ensure that the orifice exit velocity will cause no disturbance to layering. Orifice size and spacing can be adjusted to precisely control this exit velocity, accommodating any gravel design.

The Newtech underdrain's stainless steel media barrier panels permit a design that uses no gravel. The openings on the media barriers are sized to suit the media used (e.g., ES 0.90 mm, ES 0.45 mm, ES 0.25 mm, etc.).

The panels are bolted directly to both sides of each lateral, set off from the surface of the lateral to provide an intermixing zone when simultaneously air scouring and backwashing.

Filter Media

Understanding the physical characteristics of granular filter media is an essential step in understanding the overall filtration process. All the media discussed in this section are similar in nature; all rely on particle size distribution, shape, and $MnO_2(s)$ at the particle surface to accomplish oxidation/coagulation, adsorption, and filtration.

Manganese Greensand

Manganese greensand has been used for several decades in North America, often specifically for Fe and Mn removal. Manganese greensand is a purple–black granular filter medium processed from glauconite sand. The glauconite is synthetically coated with a thin layer of $MnO_2(s)$, and some particles have a definite green color, giving the material its common name. The only North American manufacturer of manganese greensand is located in New Jersey, USA.

Physical characteristics of manganese greensand. Glauconite exhibits an ion exchange capacity that allows the surface to be saturated with manganous ions. Following saturation, the glauconite is soaked in a strong oxidizing solution, which transforms the manganous ion to the insoluble $MnO_2(s)$ form. The surface coating makes up about 4.0 mg of a 1,000 mg greensand sample; about 0.4 percent of the weight of a particle of greensand is actually $MnO_2(s)$. Since $MnO_2(s)$ is black, particles not covered or only partially covered with the substance are easily identified by the greenish color of the underlying glauconite sand.

Manganese greensand has an effective size of 0.30–0.35 mm, a uniformity coefficient of less than 1.60, and a specific gravity of approximately 2.4.

Installation of manganese greensand.
1. *Initial backwash.* After placement of the greensand in a filter vessel, and prior to placing any filter coal, the greensand must be thoroughly backwashed and skimmed. Commence the backwash cycle and increase the backwash rate to full flow (called "ramping up"), and continue at full rate until the backwash effluent runs clear. The backwash water will start off black and pass through shades of gray, running light gray for some time. Depending on the depth of the manganese greensand, this initial backwashing could last from 15 to 30 min. However, this process of removing fines (i.e., manufacturing dust, particles too small to be good filtering material, and other small impurities) is critical to optimized filter runs under actual operating conditions.
2. *Greensand stratification.* After the backwash water has cleared, ramp down the backwash rate slowly to ensure that the grains of greensand stratify (as defined in the glossary). During backwashing, most of the greensand fines migrate to the surface of the filter bed, where they must be physically removed in a step known as *skimming*.
3. *Skimming greensand.* At least 2.5 cm (1 in.) per 30 cm (12 in.) in the installed greensand's depth should be removed, because extending the initial backwash cycle will not remove the majority of fines from the greensand surface. Example: If 46 cm (18 in.) of greensand has been installed, at least 4 cm (1.5 in.) should be skimmed, or removed from the surface. (Another term used to describe the physical removal of fines is *undercutting*.) Skimming is best accomplished with a flat-mouthed shovel.

The skimming step is critical, because failure to remove the fines will result in shortened filter runs. Left in the filter, the fines form a very dense layer at the top of the greensand. In a relatively short time, this dense layer becomes even denser with accumulation of turbidity from the raw water, oxidized iron, agglomerated iron and manganese, and other fine particulates. As its density increases, this layer restricts the flow of water and builds up head pressure. Backwashing becomes necessary long before the rest of the bed's capacity is reached, or it leads to a pressure rise that can fracture the glauconite and the $MnO_2(s)$ coating.

Skimming granular filter media (including manganese greensand) requires the use of the following equipment:

1. *Plywood.* Small pieces of plywood (about 60 cm x 60 cm [24 in. x 24 in.]) should be placed on the top of the media to avoid walking directly on the bed's surface and sinking down into it.
2. *Flat-mouthed shovel.* A scraping motion can be used to draw the surface material to a central location, from which it is then removed. As an alternative, the shovel can be used to undercut the surface, removing material from about 180 cm^2 (29 in.^{2}) of area per scoop.
3. *Bucket.* The easiest way to dispose of the fines is to place each scoop into a plastic bucket, then periodically dump the bucket into another container on the WTP floor for final disposal.

Conditioning manganese greensand. Virgin greensand (i.e., new and unused material) is not shipped in a regenerated form. Therefore, it must be conditioned prior to being put into service. The manufacturer recommends soaking the bed for not less than 1 h in a solution containing about 60 g (2 oz) of potassium permanganate ($KMnO_4$) per 28,320 cm^3 (1 ft^3) of media.

Greensand can also be regenerated using chlorine. Typically, a new bed is soaked in a solution containing about 100 mg/L Cl_2 for several hours.

Manganese greensand operating modes. The typical operating mode for manganese greensand treatment involves preoxidation followed by filtration. In this sequence, called continuous regeneration, a strong oxidant such as $KMnO_4$ is added ahead of the greensand filter. In theory, by continuous preoxidation of both the Fe and Mn to an insoluble particulate form, both the insoluble metals are removed by physical filtration/straining action through the greensand media. Any Fe or Mn that is not oxidized (perhaps due to underfeeding of the oxidant) is adsorbed onto the $MnO_2(s)$ surface on the greensand particles.

Operating in this mode, a filter bed accumulates solids, which are then removed during the backwash cycle. The inherent stickiness of most oxidized Fe and Mn compounds makes removal with backwash water alone a difficult task. Backwash assisted by air scour is recommended by the manufacturer, so that the bed is kept as clean as possible. Concerns about removing the medium's $MnO_2(s)$ coating by excessive air scouring could be alleviated by limiting air scour to low rates, short periods, infrequent intervals, or a combination of the three. Keeping the media clean is the key to a sustainable filtration process and extended media life.

A second operating mode for manganese greensand is oxidized iron filtration and manganese removal by adsorption. Typically, raw water is aerated to oxidize the iron, the major portion of which is then removed in the coal layer that tops the greensand. The manganese in solution remains until it is adsorbed onto the fully regenerated $MnO_2(s)$ coating. This sequence is called intermittent regeneration. Formulas have been developed to calculate appropriate filter run lengths, but a practical guideline calls for backwashing and regeneration when the manganese level in the filter effluent reaches 0.05 mg/L. Operators interested in optimum water quality end filter runs at some point before 0.05 mg/L Mn is reached in the filter effluent.

Intermittent regeneration is preferred where raw water constituents interfere with the preoxidation/filtration process. The specific nature of these interferences is not fully understood. Difficulties that appear in one region do not appear to cause problems in other areas.

In some regions, levels of organic carbons above about 2 mg/L in groundwater are almost always accompanied by both ammonia and hydrogen sulfide. This combination frequently presents difficulties for efforts to remove Fe and/or Mn to satisfactory levels. Are the Fe and Mn organically complexed in such situations? If they behave as though they are, the question is academic. Pilot testing should guide the choice of a removal process that works, rather than devoting resources to nailing down the precise chemical reason for the removal difficulty.

A third mode of operation for manganese greensand filtration involves oxidation of iron by aeration, followed by a chlorine feed at least sufficient to continuously regenerate the $MnO_2(s)$ surfaces, on which the Mn is then removed by adsorption. In this mode, sufficient chlorine can be fed to take care of both regeneration and other chemical demands, while providing the necessary free chlorine residual in the filter effluent. Typically, dosing that gives a free chlorine residual in the filter effluent of 0.5 mg/L results in continuous regeneration of the $MnO_2(s)$ surfaces.

Regeneration of manganese greensand. In any operating mode, to continue removing Fe and Mn, manganese greensand requires regeneration. The manufacturer recommends continuous regeneration for well waters where iron removal is the main objective with or without the presence of manganese. This method involves feeding sufficient $KMnO_4$ and/or Cl_2 to satisfy all chemical demands, including regeneration of sites on the $MnO_2(s)$ coating occupied by adsorbed iron and/or manganese.

Actual filter audits have demonstrated a need for careful dosing. Effluents from some filters receiving slightly pink water have contained higher levels of manganese than were found in the raw water. This effect resulted from $MnO_2(s)$ surfaces coated with oxidized iron, scale formation on the greensand particles, or a high percentage of manganese greensand particles backwashed away as the grains were replaced by heavier mud balls double the size of greensand (which have the appearance of manganese greensand when wet). All three conditions can drastically reduce $KMnO_4$ demand. In other words, a slight pink feed provided too much regenerant for a filter bed with a very low regenerant demand. Because $KMnO_4$ is about 35 percent manganese, any overfeed of it may elevate manganese residuals in the finished water.

For Cl_2 dosing, the rule of thumb is to prefeed enough to satisfy all chemical demands and give a free chlorine residual in the water leaving the treatment plant adequate to comply with regulatory standards.

The greensand manufacturer recommends intermittent regeneration for well waters needing treatment for Mn alone or for Mn with small amounts of Fe. Briefly, intermittent regeneration involves addition of a predetermined amount of $KMnO_4$ or Cl_2 in solution applied to the manganese greensand bed after a specified quantity of water has been treated. The company's recommended dosage is about 60 g (2 oz) of $KMnO_4$ per 28,320 cm^3 (1 ft^3) of manganese greensand. A Cl_2 solution of about 100 mg/L free chlorine can also be used.

These rules of thumb do not reflect a relationship between the regenerant used and the amount of Fe and Mn accumulated in the bed. Some formulas take account of the molecular weights of the metals to be removed and the oxidants to be applied, allowing calculation of amounts of oxidant for specific quantities of water treated and metals removed, assuming a fully regenerated bed at the start of the filter run and the absence of other demands on the oxidant of choice. These formulas are best applied according to site-specific considerations derived from actual examples, since the results are then often weighted by other raw-water considerations or oxidant use upstream of the filter. Consult your treatment specialist or process consultant if these calculations are required.

A 1991 research project by led William R. Knocke reached the following conclusion:

(1) The sorption of Mn(II) by $MnO_x(s)$-coated filter media is very rapid.Both the sorption kinetics and sorption capacity increase with increasing solution pH or surface $MnO_x(s)$ concentration or both.

(2) In the absence of a filter-applied oxidant, Mn(II) removal is by adsorption alone. There was no evidence to substantiate any auto-oxidative reaction between the sorbed Mn(II) and the $MnO_x(s)$ surface over the pH range examined. . . .

(3) When free chlorine is present, the oxide surface is continually regenerated, promoting efficient Mn(II) removal over extended periods of time. This means that treatment facilities that practice prefilter chlorination are maintaining these oxide surfaces in a viable state for continuous Mn(II) removal. Conversely, treatment facilities that decrease or totally eliminate prefilter chlorination may subsequently be faced with elevated Mn(II) concentrations in the effluent waters (Knocke, Occiano, and Hungate 1991).

Topping coal. The use of a filter coal (anthracite) layer on top of a manganese greensand bed is often a critical step in reducing the solids load (i.e., filtering out particulate matter from pretreated water), promoting optimum filter run lengths. The effective size, uniformity coefficient, and specific gravity of the topping coal are important choices. Poor combinations result in intermixing of the two media. A higher degree of intermixing reduces the capability of each medium to perform its intended function.

A filter coal layer over manganese greensand gives two benefits. The first results because on average the coal particles are about three times larger than the greensand particles (0.9 mm compared to 0.3 mm). Because of this difference, the spaces between the coal particles are also significantly greater, so they can hold a greater volume of filtered material than the manganese greensand can hold.

The space between particles of granular filter media is referred to as its *void area* (although some use the term *bed porosity*). The amount of filtered material needed to fill those void areas is known as the filter medium's *solids holding capacity*. Coal can hold much more filtered material than manganese greensand before the flow of water becomes restricted. Typically, filter coal with an ES of 0.7 to 0.8 mm has a void volume of approximately 60 percent. A volume of manganese greensand, on the other hand, can hold much less than the same amount of a compatible filter coal before the flow is restricted. Typically, the void volume of manganese greensand is 35 to 40 percent, according to the manufacturer. Without a coal layer, the restricted flow through the manganese greensand bed can build up pressures high enough to fracture the manganese greensand particles and their $MnO_2(s)$ coating.

The second benefit of a coal layer results because the void area in the coal provides a place for oxidized Fe and Mn to flocculate (i.e., bunch up). Pretreatment of the raw water aids filtration. For example, Cl_2 or $KMnO_4$ is added to oxidize the iron. As the iron goes through a chemical change, it usually forms solid particles that join together into bunches big enough to get caught between the particles of filter media. As the oxidized iron particles jostle through the coal layer, they collide with each other, stick together and to the coal particles, where they are trapped. This description simplifies removal of particulate.

If the oxidized iron is permitted to flow down into the manganese greensand layer, it tends to coat the particles with iron oxides over time. No further manganese can be adsorbed by the $MnO_2(s)$ at these locations, and the manganese greensand gradually loses its ability to remove manganese.

Some exceptions complicate filtration. Some iron species under certain oxidation conditions remain fine enough to sift right through the filter bed. Some of the possible solutions are increasing detention time (i.e., the length of time the oxidizing chemical is in contact with the iron before the water reaches the filter), adding a flocculant (for example, alum) following oxidation, and/or adding a polymer (sometimes referred to as a filter aid) before filtration.

Despite variations, a properly designed coal layer is an important tool for keeping the manganese greensand clean and extending its useful life. Proper design of the filter coal layer must ensure the optimum hydraulic backwash rate, since the rate for the filter coal is slightly less than the optimum rate for manganese greensand. Since greensand is produced by only one manufacturer, operators can rely on a predictable consistency of particle size and specific gravity, which aids in calculations of appropriate backwash rates. The same is not true for topping coal. Specific gravities of coal samples can vary from 1.3 to 1.8, and uniformity coefficients can vary from 1.3 to 2.0, depending on the source of supply. An optimum design includes topping coal as coarse as possible while retaining the ability to fluidize at a backwash rate just below the fluidization rate for the manganese greensand. To achieve this balance, coal with the right effective size, uniformity coefficient, and specific gravity must be chosen.

Experience suggests the following general guidelines:

1. Both the glauconite mineral sand grains and the $MnO_2(s)$ can be fractured at pressure differentials over 55 kPa (8 psi).

2. The fine particle size distribution of manganese greensand and the typical depth installed in many filters results in a high clean bed head loss. This means that water is much more reluctant to flow, even through clean manganese greensand, than through typical sand and/or coal, which have larger particle sizes. To achieve acceptable filter run lengths, a clean bed is important, which means only minimum backwash flow maldistribution can be tolerated and the backwash volume must meet specifications. Failure on either count results in reduced life of the manganese greensand.

3. If pressure differentials result in particle fracturing, fines are created and accumulate on the surface of the bed. This accumulation in turn reduces filter run lengths by raising clean bed head loss. If the production of new fines resulting from excessive pressure differentials is not identified and corrected, the cycle will repeat itself until replacement of the manganese greensand is necessary.

4. The $MnO_2(s)$ surface coating is also susceptible to attrition and surface iron fouling. Careful precautions are needed to maintain this surface coating in a clean condition, or its adsorptive capacities will be reduced or lost.

5. Some worry about attrition resulting from repeated air scouring of manganese greensand, although no available documentation substantiates the claim. The manufacturer's literature (Inversand, n.d.) recommends air/water scour as an option, using 4.1–10.2 $(L/s)/m^2$ (0.8–2.0 $SCFM/ft^2$) of air flow with a simultaneous treated-water backwash at 9.8–12.2 m/h (4–5 gpm/ft^2). In practical applications, an air scour rate of 10.2 $(L/s)/m^2$ (2 $SCFM/ft^2$) or more is required to generate the kind of vigorous action needed to loosen filtered particulate.

6. In order to maintain adsorptive capacity in manganese greensand, the bed must be regenerated, either continuously or intermittently following a backwash cycle. Once the $MnO_2(s)$ coating on each individual particle of glauconite has adsorbed all the soluble Fe and Mn it can, the bed acts only as a strainer. Regeneration results from bringing an oxidant into contact with the manganese adsorbed, which changes its chemical character to a form that can be removed by backwashing, perhaps with air scour. The same removal mechanism applies to adsorbed iron. Cl_2 and $KMnO_4$ are commonly used regenerants.

Birm™(Burgess Iron Removal Method)

The Burgess Iron Removal Method (Birm)™ is another filtration technology. It has been used mostly in point-of-use applications, with only a few applications in municipal, commercial, and industrial water treatment.*

Physical characteristics. The Birm™ medium works through an $MnO_2(s)$ catalyst impregnating aluminum silicate sand. The base material is treated with manganous salts to near saturation, and the manganous ion is then oxidized to $MnO_2(s)$ with permanganate, a process similar to the one used to manufacture manganese greensand. While manganese greensand weighs 38.5 kg/0.02 m^3 (85 lb/ft^3), the Birm™ medium weighs only 20 to 23 kg/0.02 m^3 (46 to 50 lb/ft^3). It is also lighter than anthrasand and pyrolusite. Due to the medium's light weight, backwash rates require strict control. Further, its low specific gravity prevents the use of a coal cap.

Birm™operating mode. The Birm™ process data sheet recommends air as the only oxidant to be used ahead of the filter. In most common groundwaters, dissolved Fe and Mn occur in the divalent ferrous and manganous states due to the presence of free CO_2. The medium acts as a catalyst between the dissolved oxygen and the soluble Fe and Mn compounds, enhancing this reaction and producing ferric and manganic hydroxides that precipitate to filterable form. After backwashing to remove the precipitate from the media, the filter is again ready for service flow. Acting as a catalyst (defined in the glossary), the Birm™ medium is not consumed and requires no regeneration.

General information about Birm™. The main concern for the success of Birm™ is the operating recommendation to use air as the only oxidant. Experience clearly demonstrates that the adsorptive and oxidative capacity of the medium's $MnO_2(s)$ surface becomes exhausted without continuous or intermittent feeding of some regenerating oxidant.

Although the medium can be regenerated during backwash using Cl_2, a strong Cl_2 solution might affect any exposed surfaces of the base granular material. This concern arises from the manufacturer's caution against the use of chlorine.

Attrition studies support a prediction that vigorous air scour of the Birm™ filter would result in high attrition losses because of its low hardness factor. No manufacturer's figures are available, but a comparison of the hardness of the Birm™ medium to that of manganese greensand suggests that fracturing of Birm™ particles would occur at pressure differentials in excess of 55 kPa (8 psi).

Birm™ is a registered trademark of the Clack Corporation, located in Wisconsin. A 1988 spec sheet issued by Clack offers the following advice:

---

*The Burgess Company was well-known in decades past as a maker of car batteries.

Birm™ acts as an insoluble catalyst to enhance the reaction between dissolved oxygen (DO) and the iron compounds. . . . [The reaction] produces ferric hydroxide which precipitates and may be easily filtered. . . . [The medium] is easily cleaned by backwashing to remove the precipitate. . . . When using Birm™ for iron removal, it is necessary that the water contain no oil or hydrogen sulfide (H_2S), organic matter (TOC) not exceed 4–5 ppm, the DO (i.e., dissolved oxygen) content equal at least 15% of the iron content with a pH of 6.8 or more. . . . A water having a low DO level may be pretreated by aeration. Chlorination greatly reduces Birm's™ activity and dosages should be held at a minimum. . . . Clack's Birm™ may also be used for manganese removal. . . . In these applications the water should have a pH of 8.0–9.0 for best results. If the water contains iron, the pH should be below 8.5. High pH conditions may cause the formation of colloidal iron which is very difficult to filter out.

Obviously, certain characteristics of raw water must be known before treatment using Birm™. Levels of DO, TOC, H_2S, and pH are critical, since they greatly influence the effectiveness of Birm™.

Anthrasand

Anthrasand is another filtration medium similar to greensand. A base material of standard anthracite coal and silica sand, sized in a conventional dual-media configuration, is coated with a thin layer of $MnO_2(s)$. One basic difference between anthrasand and manganese greensand is the method of applying the $MnO_2(s)$ coating. Anthrasand is placed in the filter, where it is soaked in a manganous salt solution for a prescribed time before $KMnO_4$ is added to oxidize the manganous ion to the $MnO_2(s)$ form. This process is referred to as in-situ generated manganese dioxide.

Anthrasand operating mode. The suggested operating mode begins with dosing the raw water with $KMnO_4$ ahead of the filter to oxidize the Fe and Mn. The filter then removes it. Any unoxidized Fe and Mn is adsorbed onto the $MnO_2(s)$ surface. Sufficient $KMnO_4$ is fed to keep the $MnO_2(s)$ surface in a regenerated condition.

Anthrasand offers inherent attrition resistance because of the hardness of the base anthracite coal and silica sand. The low clean-bed head loss associated with a dual-media design is another advantage.

General information about anthrasand. Establishing a complete and fully oxidized $MnO_2(s)$ coating in situ could be a tricky problem, though. Generally, anthracite coal and silica sand do not have ion exchange properties (as does the glauconite base used for manganese greensand), nor do they have surface areas increased by many pockets or caves. Although many anthracite coal and silica sand samples have taken on complete $MnO_2(s)$ coatings naturally within filters on site, the time requirements of the process vary, taking years in some cases and only weeks in others. Very little data is currently available about the factors that influence the rate at which an $MnO_2(s)$ coatings can be generated. Research by scientists such as William R. Knocke is continuing in this field.

Another mystery is why air scour seems to have little adverse affect on some sands naturally coated with $MnO_2(s)$, while air scour removes flakes of what appears to be $MnO_2(s)$ from sands in other filters. Close monitoring of backwash waters is encouraged where an anthrasand filter undergoes air scour to determine whether or not flaking is occurring.

A decision to use anthrasand, or any other medium, is predicated on the nature and treatability of the raw water. Thorough pilot testing should be completed prior to the final design to allow evaluation and development of the most efficient possible removal process.

Pyrolusite

Pyrolusite is the mineral term for naturally occurring $MnO_2(s)$. Pyrolusite filters are relatively new to North America, but the material has been the medium of choice for Fe and Mn removal in the United Kingdom for several decades.

Physical characteristics. Pyrolusite is produced from $MnO_2(s)$ ore in the United States, Australia, Brazil, and South Africa. The ore is crushed to specific sizes needed for potable water filtration processes. The resulting particles are solid pieces of $MnO_2(s)$, eliminating the need to develop a skin of $MnO_2(s)$ on each particle, as is the case with manganese greensand, BIRM™, and anthrasand. Because pyrolusite has a specific gravity in the area of 4.0 (as compared to 2.4 for manganese greensand), air scour is necessary to keep the particles of pyrolusite scattered throughout the sand bed.

Pyrolusite operating mode. A typical media bed is a blended matrix of pyrolusite and sand. Air scour helps to return a homogenous blend, depending on the process requirements. Pyrolusite/sand blend ratios range from 10 percent to 50 percent by volume.

The amount of pyrolusite in this mix depends on the nature of the raw water and the treatment program. For a process with continuous regeneration, only a relatively small amount of pyrolusite is required to provide the adsorption sites needed for relatively small amounts of manganese in the raw water. Blends up to 50 percent pyrolusite by weight have been required in circumstances with high levels of both Fe and Mn. Appropriate pilot testing is often the best way to determine pyrolusite requirements.

A solid $MnO_2(s)$ medium survives a vigorous backwash program including air scour without the worry of surface attrition. Any surface attrition of pyrolusite only exposes a fresh $MnO_2(s)$ surface.

Typically, $KMnO_4$ is not used in a pyrolusite process, although it could be. Usually, raw water is aerated to remove unwanted gases such as hydrogen sulfide and to oxidize Fe. The Fe is then either filtered in a coal layer above the pyrolusite/sand matrix bed, or a coagulant is added to force particle size growth before filtration. The manganese is then removed by adsorption.

Where possible, sufficient chlorine is fed ahead of the filter to keep the pyrolusite in a regenerated condition. Where formation of unwanted chlorine by-products is a concern, chlorine is used intermittently to regenerate the bed following backwash.

As for any filter medium, the ability to clean the bed thoroughly is of high importance. Air scouring pyrolusite provides excellent results with no significant change to the particle size or shape.

General information about pyrolusite. Because of pyrolusite's high specific gravity, some plants cannot provide the necessary backwash flow to fluidize and adequately clean the bed. Simultaneous water backwash and air scour do the best job of cleaning the media. Many plants conduct air scour first, followed by hydraulic wash, then another short air scour to redistribute the pyrolusite in the sand bed.

Summary

Common granular filter media in use for the removal of Fe and Mn are similar in that they all rely on a surface of $MnO_2(s)$. Base materials vary, but the function of the $MnO_2(s)$ surface is often the same.

1. Every common medium has an $MnO_2(s)$ surface.
2. All $MnO_2(s)$ media can be regenerated with either Cl_2 or $KMnO_4$.
3. $MnO_2(s)$ surfaces must be kept in a clean condition to sustain the removal process. Media life is directly related to the efficiency of the backwash, the capabilities of the underdrain design, and backwash operating procedures.
4. In some cases, media life is directly related to the attrition resistance of the media particles.
5. Pilot testing should be done for any medium to ensure removal efficiencies, acceptable filter run lengths, and the adequacy of media bed cleaning mechanisms. Pilot testing should always be done prior to design and construction of a water treatment plant.

Chemically Enhanced Filtration

Adding coagulants and flocculants makes water filterable; more precisely, it makes substances to be removed from the water, in this case Fe and Mn, filterable. If Fe and Mn are oxidized into an insoluble form, the oxidized particles may still be too small to become trapped in a granular media filter. In such a case, additional chemicals are needed to agglomerate the oxidized particles into units big enough to filter.

As a rule of thumb, a granular media filter should remove suspended solids (including oxidized Fe and Mn) greater than 10 μm in size. Extensive field data indicate that a media bed properly designed and operated at optimum levels does remove most particles in the 5 to 10 μm size range. Particles smaller than 5 μm usually sift through the filter and show up as Fe and Mn residuals in the finished water.

A large variety of chemical coagulants are available today, but stripping them of their marketing frills leaves more similarities than differences. These chemicals include coagulants and flocculants.

Coagulants are chemicals to neutralize/destabilize the surface electrical charges on particulate matter, allowing flocculation to take place. Flocculants are chemicals that create flocs composed of the substances themselves and aggregations formed as the result of large numbers of collisions between coagulated particles. Flocs can be either settled out or removed by granular media filtration. Chemicals are often combined to serve the dual function of coagulation and flocculation.

For example, most impurities found in water—clays, siliceous material, organic colloids—carry negative surface charges. Aluminum salts such as alum and the polyaluminum compounds hydrolyze in water to form positively charged ionic species that neutralize the negative surface charges on particulates, allowing them to agglomerate. The aluminum reaction continues to form insoluble aluminum hydroxide, a gelatinous substance that traps the coagulated impurities. Iron salts are sometimes used as coagulants, as they too form positively charged species including ferric hydroxide.

Some water treatment processes, including lime softening, generate $CaCO_3$ precipitates that carry negative surface charges. Overfeeding chemicals may also generate magnesium hydroxide, which has a positive surface charge. If the softening operation generates very little magnesium hydroxide, then some supplemental coagulation may be required to assist in removal of precipitated $CaCO_3$.

Widely used in the water treatment industry and often very useful in Fe and Mn removal processes are organic polymers. These chemicals may function as coagulants, flocculants, or both or as floc tougheners.

Organic polymers are grouped as anionic, cationic, and nonionic polymers:

1. Anionic polymers carry negative charges. Dosages do not exceed 0.5 mg/L, and typical dosages are in the 0.1–0.3 mg/L range. Overfeeding this polymer (i.e., more than 1.0 mg/L) may result in filter media blinding, which is chemical agglomeration of the media grains themselves. In effect, excess anionic polymer tends to glue together the filter media grains.
2. Cationic polymers carry positive charges. Typical feed rates are in the 2–10 mg/L range.
3. Nonionic polymers carry no charges. They are typically used as floc tougheners. Floc generated in cold water is often fragile and light, and addition of a

nonionic polymer in small doses makes the floc more resilient to shearing (i.e., breaking apart) from mixer action, pumping, or other turbulent conditions. Once sheared, the broken down floc does not grow again to its optimal size, and it may settle too slowly and/or pass through the filter bed. Media blinding can result from overfeeding nonionic polymers.

If Fe cannot be removed from groundwater by direct filtration (i.e., oxidation/detention/granular media filtration), it may be in one of two forms, both requiring addition of a polymer prior to filtration:

1. If Fe is organically bound, it carries a negative charge. Oxidation using $O_2(aq)$ or Cl_2 may oxidize ferrous iron to its ferric form, though the ferric iron can still be bound to its carbon host and/or in colloidal form. A cationic polymer neutralizes the colloidal charge and allows flocculation to filterable size.
2. Using Cl_2 as an oxidant sometimes results in a positively charged colloidal ferric hydroxide. An anionic polymer neutralizes the colloidal charge to permit flocculation to filterable size.

Chemical polymers require cautious use because of the potential for media blinding. For a process without an air scour capability or less than optimum backwash rates, pilot testing should almost certainly be done to measure and observe the effects on the media bed of the polymer at its optimum dosage. Failure to properly assess the impact of a polymer program could drastically shorten media bed life or blind a media bed beyond recovery.

If iron breakthrough (i.e., elevated levels of Fe in the finished water) occurs fairly soon after the start of a filter run, a quick test helps to establish whether or not use of a flocculant is indicated:

1. Apply the oxidant, and permit the chemical reaction to reach its conclusion. Then test to confirm that the Fe has taken its oxidized form. A logical source of the sample would be just ahead of the filter.
2. Pass the sample through a 5 μm filter membrane, and measure the amount of iron in the filtered product. If most of the oxidized Fe passed through the 5 μm membrane, dosing an appropriate chemical coagulant/flocculant may result in units of oxidized Fe large enough to remove by granular media filtration. (If the 5 μm filter membrane removed the iron but the actual filter didn't, an audit of the filter is indicated. Possible explanations could be channeling, higher than optimal rates of filtration, and build-up of filtered particulate resulting from inappropriate backwash rates, among others.)

Addition of a chemical coagulant/flocculant should encourage agglomeration of Fe to units greater than 10 μm in size. Groundwater treatment plants using pressure filters for Fe and Mn removal require in-line static mixers immediately following the point of coagulant/flocculant injection to ensure the chemical is thoroughly mixed with the raw water.

Matching Media Sizes

Design of a media bed requires application of a formula to find a theoretical match between media sizes before loading pilot columns with media:

$$\frac{d_1}{d_2} = \left(\frac{p_2 - p}{p_1 - p} \right)^{0.667}$$

Where:

d_1 = the particle size of the first medium
d_2 = the particle size of the second medium
p_1 = the density of particles of size d_1
p_2 = the density of particles of size d_2
p = the density of the fluid (1.0 for water)

Application of this formula assumes the correct choice of the first filter particle size has been made. It also points out the need for compatible specific gravities if a filter needs large media as a top layer to accommodate extensive particulate loading. The formula does not take into account uniformity coefficients of the two media, though.

As an example, suppose someone must choose a particle size d_2 to accompany sand (d_1 = 0.45 mm, p_1 = 2.65). The required density, p_2, is 1.63, and the density of water, p, is 1.0 by definition:

$$\frac{0.45}{d_2} = \left(\frac{1.63 - 1}{2.65 - 1} \right)^{0.667}$$

d_2 = 0.85 mm, the effective size of the coal

Theoretical matched size is just that, a theoretical value. Experience in media bed design suggests combinations most likely to do the job. Regardless of experience level, pilot testing should be done to be sure that a dual- or multiple-media bed will produce the quality of water required, given the nature of the raw water and the chemical pretreatment program.

L/d Ratio

A filter bed design also requires calculation of the L/d ratio, the relationship between the size of the media and the depth of the bed. Overall criteria include:

$L/d \geq 1,000$ with polymer added as a filter aid
$L/d \geq 1,500$ without polymer addition

Where:

    L = depth of the bed, in mm

    d = effective size of the medium, in mm

Example. What is the L/d ratio for a filter bed of 508 mm (20 in.) of coal with 1 mm effective size and 305 mm (12 in.) of sand with 0.55 mm effective size?

    L/d for the coal layer = 508/1 = 508 mm (20.0 in.)

    L/d for the sand layer = 305/0.55 = 555 mm (21.9 in.)

    L/d of the filter bed = 508 + 555 = 1,063 mm (approximately 42.5 in.)

Considerations for Media Selection

Several additional factors must be addressed prior to establishing a final filter media design. These factors include, but are not limited to:

1. Nature of the raw water (i.e., level of contaminants)
2. Required effluent quality
3. Type of process (i.e., continuous or intermittent regeneration)
4. Filtration qualities of the raw water
5. Available filtration equipment to do the job (i.e., vessels, piping, valves, pumps, instrumentation, etc.)

Detailed pilot studies should be conducted before finalizing a design or upgrade of any filtration process.

Appropriate Backwash Rates

An appropriate backwash rate is defined as the flow of water through a filter bed that achieves the optimum cleaning action on the granular filter media. The point at which optimum cleaning is reached is defined as maximum hydraulic shear velocity. In other words, if water flows past a grain of filter medium at exactly the right speed (velocity)—not too fast and not too slow—it has reached its maximum shear capability (i.e., the highest possible capability to separate filtered particulate from a grain of filter medium). Maximum hydraulic shear occurs at lower velocities in cold water than in water of higher temperature, because temperature alters water's viscosity. (See the glossary for definitions.) Hence, optimum backwash rates, especially for surface waters, are lower in the winter months than in the heat of summer.

In the old days, many manufacturers of filter media specified backwash rates on the basis of how much the filter bed expanded when backwash water was applied. A specifications sheet would say something like, "backwash until the bed has expanded 35 to 50 percent." Give or take 15 percent is a wide range! Further, operators struggled to determine when beds had expanded, by say 35 percent, while water was flowing through the beds, especially in closed pressure filters from which the backwash water exited the filter through lines in the tops of the vessels.

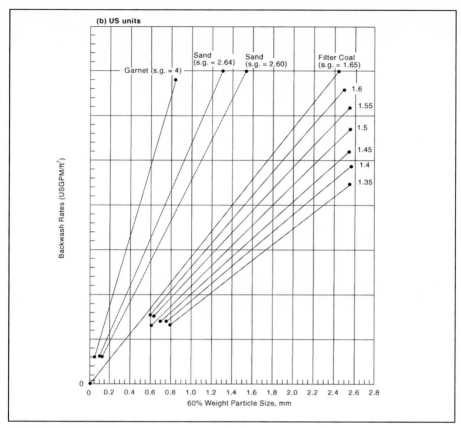

(b) US units

Source: Anthratech Western Inc.

Figure 6-14 Appropriate backwash rates for various filter media at 20°C (68°F)

To calculate precise backwash rates, the effective size, uniformity coefficient, and specific gravity of the filter medium must be known. See Figure 6-14.

If the filter bed is composed of two or more media, the backwash must effectively fluidize all media at approximately the same rate of flow. For example, if manganese greensand fluidizes at, say 30 m/h (12.5 gpm/ft²), topping coal should be chosen with an effective size, uniformity coefficient, and specific gravity that permit it to fluidize at slightly below the greensand rate. This practice keeps the coal together in a layer above the greensand. Because the coal should be as coarse as possible, choosing a coal with too high a specific gravity risks intermixing of the coal and greensand.

Trial calculations using various UC and SG values show that both obviously play important parts. Choices of ES and UC are severely limited if the only coal available is anthracite with a high specific gravity (e.g., SG 1.65 or higher). All other things

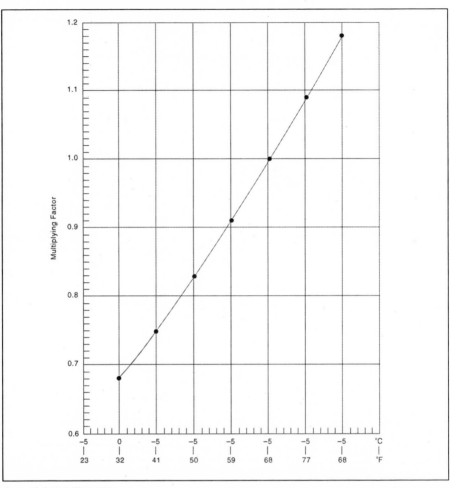

Source: Anthratech Western Inc.

Figure 6-15 Appropriate backwash rates—temperature correction factor

being equal, topping coal SG can dramatically affect the solids-holding capacity of the filter bed and therefore the length of filter runs. Temperature also affects backwash rates, as Figure 6-15 shows.

Appropriate Air Scour Rates

All other things being equal, the effectiveness of air scour in cleaning filter beds cannot be disputed. Discussion continues in the industry, however, concerning appropriate rates.

The manufacturer of manganese greensand recommends air/water scour as an option, using a rate of 4.1–10.2 (L/sec)/m^2 [0.8–2.0 SCFM/ft^2] with a simultaneous treated water backwash at 9.8–12.2 m/h (4–5 gpm/ft^2). This recommendation leaves a substantial spread between the high and low limits. The company offers no guidance for when 4.1 (L/sec)/m^2 is appropriate, and under what circumstances the rate should be higher, but not exceed 10.2 (L/sec)/m^2.

Extensive empirical data supports air scour rates between 12.7 and 20.3 (L/sec)/m^2 [2.5 and 4.0 SCFM/ft^2]. Rates at the high end of the scale are required with filtering material of higher SG than common filter sand and filter beds over about 90 cm (about 36 in.) deep. As a rule of thumb, appropriate rates fall about 15.2–17.8 (L/sec)/m^2 at 28–40 kPa (3–3.5 SCFM/ft^2 at 4–6 psi). Common filter coals and sands, including manganese greensand, do not exhibit high rates of attrition (i.e., particle size reduction from wear during air scour) at these rates. The losses are low enough to cause no practical effect on filter bed life expectancy.

Following these guidelines is a challenge, since equipment provides no easy way to measure the air input applied. In most cases, the manufacturer's information provided on plates on the blower and electric motor driving it are used to calculate blower output. (Note: Air compressors are not suitable for air scour, because they produce low volumes of air at high pressures.)

Although simultaneous air/water backwash is the most effective cleaning combination, many filters (especially pressure filters) do not lend themselves to this method. Normally a filter is drained down 30–60 cm (a foot or two) below the backwash water outlet, air scour is applied then turned off, and appropriate backwash flows are applied until the water runs clear from the outlet. If appropriate backwash flows are not possible because of piping sizes, pump capacities, or design deficiencies, extending the period of air scour helps to move filtered particulate near the surface of the filter bed. The low flows may then be able to transport the filtered particulate to waste even though optimum hydraulic shear is not achieved.

Calculating Backwash Rates

Measuring backwash rates is another challenge in plants with rudimentary backwash flow indicators, indicators that haven't worked for years, or designs that never included backwash flow indicators. Table 6-1 shows how a metre-stick and stop watch can be used to measure the rate at which backwash water rises in a filter and how that rise is then interpreted as a m/h (gal/ft^2) number.

The simple, fool-proof method includes seven steps:

1. Drain the filter so the surface of the media bed is visible. The water level must be below the surface of the bed by 2 to 5 cm (1 to 2 in.).
2. Hold the metre-stick so the bottom is just above the media bed or just touching it. Attaching the metre-stick to some fixture (say, clamping it to a backwash trough) will likely improve the accuracy of the reading. If the bed is loose and fluidizes easily, a carpenter's tape measure can be used.
3. Have an assistant turn on the backwash flow to full rate (the usual rate used in the plant). If starting full flow from a dead stop will create high pressure

stresses on piping and valves (known as water hammer), slowly ramp up the backwash flow. This requirement means the filter should be completely drained first, so that full backwash flow can be achieved by the time the water floods the surface of the media bed and reaches the bottom of the metre-stick or other measuring device.

4. Start the stop watch as soon as the water rises to the bottom end of the metre-stick.

5. Stop the stop watch when the water has risen 61 cm (24 in.). The watch can be stopped after, say, 30.5 cm (12 in.), or at any other distance, but 61 cm gives a good calculation base. As an alternative, record how high the water rises up the metre-stick in 1 min.

6. Compare the readings to data in Table 6-1. For example, a rise rate of 61 cm in 1 min gives a backwash flow per square metre of 36.7 m/h (15 gpm/ft^2, or 12.5 Imperial gpm/ft^2).

7. Calculate the flow per square metre (ft^2) of filter bed area.

Example. Backwash flow rises 61 cm in 1 min in a pressure filter with a 152-cm diameter. What is the total backwash flow in m/h? Table 6-1 shows that 61 cm of rise in 1 min translates into 36.71 m/h. A 152 cm diameter filter has a surface area of approximately 1.81 m^2 (Area = πr^2 = 3.1416 × 76 cm × 76 cm). Therefore, the total backwash flow is 36.71 × 1.81 = 66.45 m/h.

Table 6-1 Backwash flows based on vertical rise rates

Vertical Rise, cm/min	Vertical Rise, in./min	Backflow Rate, m/h (m^3/m^2/hr)	Backflow Rate, gpm/ft^2	Backflow Rate, Imp gpm/ft^2
2.5	1	1.52	0.62	0.52
5.1	2	3.10	1.25	1.04
7.6	3	4.58	1.87	1.56
10.2	4	6.12	2.50	2.08
15.2	6	9.15	3.74	3.12
20.3	8	12.19	4.98	4.15
25.4	10	15.27	6.24	5.20
30.5	12	18.21	7.44	6.20
40.6	16	24.38	9.96	8.30
50.8	20	30.55	12.48	10.40
61.0	24	36.71	15.00	12.50
71.1	28	42.83	17.50	14.60
81.3	32	48.95	20.00	16.60
91.4	36	54.92	22.44	18.70
101.6	40	61.09	24.96	20.80

Chapter 7

Alternate Technologies

Iron and Manganese Removal by Softening

Iron and manganese can be effectively removed, along with hardness, either in a zeolite water softener or as part of a lime or lime/soda ash softening process.

Removal by Zeolite Softener

A zeolite water softener (see the glossary) removes up to several milligrams per litre (mg/L, sometimes referred to as ppm) of Fe and Mn. A typical process is shown in Figure 7-1. The flow to be softened is pumped into a pressure filter containing a chosen zeolite, which removes hardness as well as free cations of Fe and Mn. The ions are bound to the zeolite medium, which eventually uses up all its cation exchange capacity. Regeneration is then achieved by backwashing the zeolite medium with a brine solution, usually prepared with sodium chloride (NaCl). The Fe and Mn ions leave the filter in the backwash waste flow, sodium ions replacing them on the medium. Following regeneration, the zeolite medium is rinsed with clean water before being returned to service.

An important restriction requires that no oxidants be added to the water on its way to the softener. Otherwise, chemical oxidation of the Fe and Mn may occur, plugging the zeolite medium or coating it with oxidation products.

Removal by Lime/Soda Ash Softening

When water is softened using lime or lime/soda ash softening processes, a beneficial side effect is virtually complete removal of Fe and Mn. Adding lime or soda ash raises the water pH to about 10 if only calcium is to be removed or about pH 11 if the ferrous and manganous ions are to be precipitated as iron and manganese hydroxides. This precipitate is removed along with the mainly calcium carbonate sludge. In this process, the Fe and Mn are oxidized by O_2 to Fe^{+3} and Mn^{+4} (though the Mn could also go to Mn^{+3}) before the softening step. Usually, insignificant levels of Fe and Mn remain, and no additional Fe and Mn removal treatment is needed.

The softening process usually occurs within a solids contact type clarifier. Significant levels of noncarbonate hardness often call for a two-stage softening process. Extensive chemical handling and dosing systems are required, and large volumes of sludge are produced. Consequently, while the process is extremely effective for removing Fe and Mn, it is not an economic way to achieve this goal, unless the water must also be softened.

Sequestration of Iron and Manganese

Sequestration (see the glossary) is often used to prevent oxidation of Fe and Mn within the distribution system. This treatment helps to keep the system clean and the water free from objectionable red and/or black color and turbidity.

The sequestration process does not remove either Fe or Mn, but it reduces the color and turbidity associated with their presence. The process involves the addition of a sequestering agent, such as sodium silicate in conjunction with chlorine, or

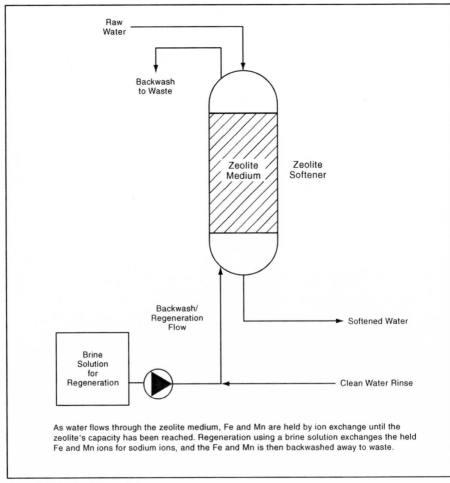

As water flows through the zeolite medium, Fe and Mn are held by ion exchange until the zeolite's capacity has been reached. Regeneration using a brine solution exchanges the held Fe and Mn ions for sodium ions, and the Fe and Mn is then backwashed away to waste.

Source: Reid Crowther & Partners Ltd.

Figure 7-1 Zeolite softening

polyphosphates in conjunction with chlorine. These agents delay the formation of significant color and turbidity caused by the gradual oxidation of Fe and Mn within the distribution system. Their effect is temporary, but it usually lasts longer than the water's stay in the distribution system.

A chemical supplier should be consulted on the effective in-system life of particular products. Some small plants with several days' storage capacity may find that a sequestering product's effect lasts too short a time, especially during minimal consumption periods such as the dead of winter.

In Fe sequestration with sodium silicate, chlorine is first added to oxidize the iron, changing it from the ferrous form to the colloidal ferric form. (This oxidation process is not a contradiction of the process goal, mentioned earlier, of preventing oxidation, since this statement refers to a process using only sodium silicate.) Sodium silicate or a polyphosphate is then added to stabilize the colloid, keeping it too small to cause any apparent color or turbidity. If color and/or turbidity appear, the chlorine is being fed too far ahead of the sequestering agent.

Bench or pilot testing should be done to determine how quickly color problems form, so that detention time can be calculated and the appropriate injection points chosen. When treating water with a sodium polyphosphate, the polyphosphate is generally added prior to the Cl_2.

Manganese sequestration works in a different way than iron sequestration. Instead of oxidizing the Mn, as is done with Fe, sequestering agents delay the oxidizing effect of chlorine on Mn in solution. Instead of producing small precipitates of Mn, the sequestrants delay production of any precipitates of Mn at all.

As a rule of thumb, sequestration of Fe and Mn should be undertaken only following on-site pilot testing and a cost–benefit analysis, or comparison with a similar system. Usually, sequestration attempts to deal with the small amounts of Fe and Mn left following other removal processes (such as oxidation/filtration or adsorption) and to bring about other benefits. The cost to sequester several mg/L of Fe and anything over 0.5 mg/L of Mn in water with high hardness should be compared to processes for Fe and Mn removal before sequestration.

Sequestration Agents

A sequestration agent is actually a chemical. Usually, either sodium silicate or polyphosphates are used to sequester Fe, and polyphosphates are used to sequester Mn. Many modern polyphosphates are long-chain linear phosphates; the arrangement of the chemical's molecules in a long chain helps them to attach to the material to be sequestered and hold it in that form for a certain time, or until some other factor breaks the sequestration links.

Sodium silicate is available in liquid form, with a silicate content between 20 and 36 percent, measured as SiO_2. The ratio of SiO_2 to Na_2O is important, since a rising ratio provides an increasingly viscous solution of increasing molecular weight. The pH of the solution is usually between 11 and 13. In a soft, low-alkalinity water supply, sodium silicate can raise the pH of the water.

A polyphosphate is a condensed (molecularly dehydrated) phosphate. In dry form, the product resembles crushed glass or a fine powder. Normally, it is available as a solution with PO_4 content ranging from 20 to 40 percent. The shelf life of the product is limited; it tends to revert to the orthophosphate form, particularly under high temperatures.

Sequestration of iron by chlorine and sodium silicate. The sequestering agent in this process should be added after the chlorine has passed a few minutes of detention time. If too much time is allowed for the chlorine to react, the colloids of oxidized iron grow, increasing the water's color and turbidity. If too little time is allowed, formation of oxidized Fe colloids may not finish. Pilot testing is recommended to establish the optimal detention time.

Sequestration shows irregular effects of pH on process effectiveness. No clear pattern determines whether low or high pH works best, so bench or pilot testing is necessary to find the best pH for each raw water.

Generally, 5 to 25 mg/L of sodium silicate, measured as SiO_2, will sequester 2 to 3 mg/L of Fe in the absence of calcium. However, increasing levels of calcium and magnesium, along with the hardness they cause, affect the performance of sodium silicate, requiring rising dosages. Typically, a water with 120 mg/L of calcium at pH 7 requires approximately 40 mg/L of sodium silicate (as SiO_2) to sequester 2 mg/L of Fe.

Sequestration of iron by chlorine and polyphosphates. Significant differences have been found in the effectiveness of various types of polyphosphates. Pilot testing with a variety of types is important to determine the most cost-effective approach. Some types have much longer shelf lives than others. Some are stable at up to $100°C$ for extended periods, while still others can provide benefits within distribution systems other than just Fe and Mn sequestration.

Polyphosphates produce the best results at low pH levels, but again, calcium and magnesium (as water hardness) affect the process. A possible reaction between the calcium/magnesium and the polyphosphate can create a precipitate, increasing turbidity in the water. Consequently, dosing of polyphosphate must not exceed the amount required for adequate sequestration. Also, polyphosphates should be added a few minutes before (upstream of) the point of chlorine addition.

Planning requires total hardness figures for a community's water supply, and product data sheets should provide references for feed rates under varying raw-water conditions. Add supplier assistance plus on-site pilot testing, and the cost–benefit relationship can be determined.

Typically, 2 to 4 mg/L of polyphosphate (measured as PO_4) is required to sequester 1 mg/L of Fe.

Sequestration of manganese. Polyphosphates, in conjunction with chlorine, appear to work best for sequestering Mn. Little success has come from attempts to use sodium silicate.

The polyphosphate should be added after the chlorine dosing point. Because the oxidation rate for Mn is much slower than that for Fe, the amount of detention time is not a critical factor.

Again, calcium and magnesium hardness reduces the effectiveness of the process, raising the required dosage of polyphosphates. This condition also causes the production of calcium/magnesium polyphosphate precipitates, which increase the turbidity of the water.

Most types of polyphosphates, for similar dosages (measured as PO$_4$) appear to offer similar performance in sequestering either iron or manganese. Use of strong bleaches or other chlorine-based household cleaners may overpower the sequestering agent once the water is used in the home, however. This situation could cause the Mn to precipitate and stain laundry. Although this is not a common problem, operators should be aware it can happen.

Sequestration Equipment

The equipment used for sequestration is a very simple set of chemical pumps very similar to those used for hypochlorite dosing. The sequestering agent is usually delivered to the water treatment plant and stored as a liquid. Sodium silicate can be dosed in either diluted or undiluted form. Polyphosphate should be dosed undiluted, as the product comes from the manufacturer.

For a water plant or well output that is adjusted manually, a simple, adjustable-stroke chemical dosing pump is used to add the agent to the raw water. The stroke length is manually adjusted to match the well or plant flow rate. Where the plant or well output constantly changes, a "paced-to-flow" chemical pump should be used.

Some important principles guide the choice of a location for adding any sequestering agents:

1. The dosing point must allow the necessary detention time based on a calculated distance from the chlorine addition point.
2. The point must provide good and rapid mixing of the agent. Consider the installation of a static mixer located immediately after chemical injection (dosing) points.

Advantages and Disadvantages of Sequestration

Advantages of sequestration.
- It helps keep the distribution system clean by preventing precipitation of Fe and Mn, which leads to deposits.
- It requires little capital investment beyond small chemical-injection systems similar to those used for hypochlorite dosing.
- It is easy to operate and maintain.
- It is particularly suitable for small systems, such as one in which several well sources all feed separately into the distribution system. For such a system, a

separate chemical dosing facility can be easily installed at each well head, avoiding the need for an expensive central treatment plant. (This type of application has some limitations, though.)

- Sequestration produces no waste sludge.
- Some blends of orthophosphates and polyphosphates have also proved effective in limiting the amount of copper and lead leaching from pipes within consumers' homes.

Disadvantages of sequestration.
- It is a temporary solution to a problem with Fe and Mn. Usually the effect of the sequestering agent lasts longer than the time water stays in the distribution system, but if water must be stored for more than a few days, problems associated with Fe and Mn may still arise.
- Even following sequestration, Fe may precipitate out in domestic water heaters, although this result does not affect the color and turbidity of the water. If water heaters are not routinely flushed, "slugs" of precipitated iron may pass through and cause consumer complaints.
- High levels of calcium and magnesium (i.e., water hardness) inhibit the effectiveness of the process.
- The addition of polyphosphate increases the nutrient level of the water. Unless an adequate chlorine residual is maintained throughout the system, these nutrients may support regrowth of bacteria, causing elevated HPC counts. Some polyphosphate manufacturers claim their formulas have overcome this unwanted side effect. Another potential unwanted side effect could result if levels of phosphorous in a community's wastewater were to rise to a level that exceeds a phosphorous discharge limit.
- Sodium silicate increases the sodium content of the water. This effect may become a concern in dosing waters with already high sodium levels, in which case other sequestering agents should be used. Individuals with certain heart conditions are routinely advised to drink low-sodium water.
- Depending on the raw-water quality and the condition of the distribution system, the required sequestrant feed rate may prove too high to make economic sense. As part of a pilot study to determine the practicality of using sequestrants, costs should be calculated and compared to those for other removal methods.

Biological Methods for Removing Iron and Manganese

Iron and manganese can also be removed using biological as opposed to physical/chemical means. Although biological removal processes are not common in North America, this approach has been used successfully in a number of water treatment plants in Europe, Africa, and elsewhere. However, biological treatment requires specific raw-water qualities and conditions, and not all groundwaters or surface waters can be treated economically using this technique. Where it can be used, biological treatment offers lower operating and capital costs than comparable physical/chemical processes, though, and it produces less waste product, which allows easier dewatering and disposal.

The abilities of certain types of bacteria to absorb/adsorb dissolved Fe and Mn, reducing them using enzymatic/catalytic action, has long been known. The biological treatment process encourages the growth and maintenance of large colonies of these bacteria, usually within a filter, where they can act on the dissolved Fe and Mn. The bacterial action oxidizes the Fe and Mn, and the resulting precipitates are trapped within the surrounding filter medium. The process is generally rapid, with filtration rates substantially higher than those in physical/chemical treatment processes. Success depends on creating the right environmental conditions within the filter to permit the most beneficial bacteria to develop and to maintain a strong colony. However, since different environmental conditions are required for each of the bacteria that remove Fe and Mn, treatment of a water containing both elements requires a two-stage process. Usually, an initial biofiltration stage removes Fe, and a second biofiltration stage removes Mn.

Biological Iron Removal

Many bacteria oxidize ferrous iron to the ferric form, causing it to precipitate. Bacteria commonly found to undertake this role include *Gallionella*, *Leptothrix*, *Crenothrix*, and *Siderocapsa*. Generally the bacteria oxidize the iron using either of two processes:

- intercellular oxidation (i.e., oxidation inside bacterial cells) by enzymatic action
- extracellular oxidation (i.e., oxidation outside bacterial cells) by catalytic action of excreted polymers

As mentioned previously, desired bacteria require the proper conditions to develop. Figure 7-2 shows the field of activity of iron bacteria, using pH and redox potential as the graph axes.

Generally, successful operation requires a pH level of 6.5 to 7.2, dissolved oxygen (DO) of 5 to 25 µg/L, and a temperature of 10 to 25°C. These guidelines cover a broad range of conditions. Under some circumstances, pH levels in excess of 7.2 do not hinder biological iron removal; in some instances, only very low levels of DO are required, and the temperature range may vary, depending on the type of iron bacteria used.

The raw-water must not contain disruptive amounts of compounds toxic to the bacteria, however. Toxic compounds include:

- chlorine (Cl_2) (For this reason, chlorination for water disinfection should follow completion of the biological removal processes.)
- hydrogen sulfide (less than 0.01 mg/L)
- heavy metals (for example, zinc less than 0.5 mg/L)

- ammonium nitrogen (NH$_3$)
- phosphates
- organics
- hydrocarbons (i.e., any compound containing only hydrogen and carbon, such as benzene and methane)

A typical biological iron removal process is shown in Figure 7-3. The raw water is first oxidized using direct air injection, ensuring ideal redox potential and dissolved oxygen (DO) level for bacterial growth. The process must avoid overoxidizing the raw water, or conditions appropriate for physical/chemical removal will develop, especially with pH greater than 7. This method requires precise calibration and monitoring of the oxidation process.

Bacteria can develop within both open and pressurized filters. The process forms a much more compact oxidized Fe precipitate than normally found in physical/chemical removal processes. Consequently, high filter rates in the range of

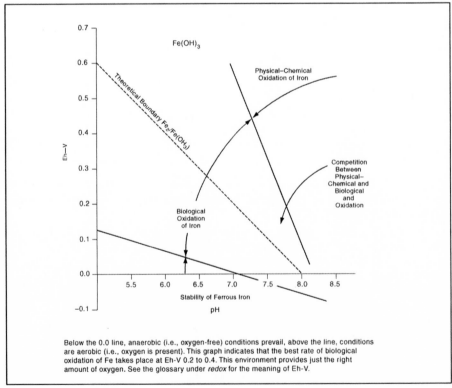

Below the 0.0 line, anaerobic (i.e., oxygen-free) conditions prevail, above the line, conditions are aerobic (i.e., oxygen is present). This graph indicates that the best rate of biological oxidation of Fe takes place at Eh-V 0.2 to 0.4. This environment provides just the right amount of oxygen. See the glossary under *redox* for the meaning of Eh-V.

Source: Reid Crowther & Partners Ltd.

Figure 7-2 Field of activity of iron bacteria

20–60 m/h (8–24 gpm/ft^2) can be used. These high rates call for a relatively coarse granular filter medium of about 2.0 to 2.2 mm effective size to limit head loss.

The filter is backwashed in the normal manner, but only with nonchlorinated backwash water. Backwash must also avoid excessive disturbance of the biofilm (i.e., the very thin layer of bacteria on each particle of the granular filter medium). Backwashing normally can use only water. Some designers have suggested that air scour should be avoided as the vigorous cleaning action tends to remove more biofilm than desirable from filter media, leading to reduced performance in subsequent filter runs. Normally, filters can be returned to service immediately following backwashing or after a prolonged shutdown without significant loss of performance.

When a new plant is started up for the first time, bacteria normally develop naturally, reaching adequate levels within 2 to 3 days in some cases and a week or so in others. Operators need not seed the filter with bacteria colonies. Other advantages of the biological process are:

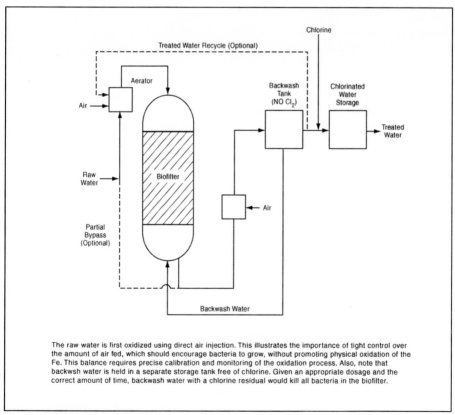

The raw water is first oxidized using direct air injection. This illustrates the importance of tight control over the amount of air fed, which should encourage bacteria to grow, without promoting physical oxidation of the Fe. This balance requires precise calibration and monitoring of the oxidation process. Also, note that backwsh water is held in a separate storage tank free of chlorine. Given an appropriate dosage and the correct amount of time, backwash water with a chlorine residual would kill all bacteria in the biofilter.

Source: Reid Crowther & Partners Ltd.

Figure 7-3 Biological removal of iron

- ease of filtering biologically reduced iron
- high production rates
- ease of dewatering sludge in the backwash water

Usually four to five times more water can be treated between backwashes using a biological process than in a conventional physical/chemical process. The precipitates of slightly hydrated iron oxides is much more compact than the precipitates formed in the physical/chemical removal processes (Voorinen et al. 1988). The dense sludge is less likely to clog filters and easier to thicken and dewater. Backwash waste can also easily be thickened by gravity settling, and the resulting sludge can be dewatered by centrifuging (spinning it in a tub until all solids gather on the wall of the tub while the water escapes) or belt pressing (forcing the sludge between two belts that press out the water).

Biological Manganese Removal

Biological removal of Mn employs a process similar to that for biological Fe removal. Both aerate the incoming raw water, both oxidize the metal (in this case Mn) through actions of bacteria, and both filter out the precipitates. However, the environmental conditions necessary to support the appropriate bacteria differ, so Mn removal must be done separately from biological Fe removal.

Bacteria active in removing Mn are *Leptothrix, Crenothrix, Siderocapsa, Siderocystis*, and *Metallogenium*. Generally, the bacteria reduce the Mn through a biocatalytic process, including intercellular oxidation by enzymatic action, adsorption of dissolved Mn at the surface of the cell membranes, and extracellular oxidation by catalytic action of excreted polymers. Mn is deposited as manganese dioxide, $MnO_2(s)$, a black precipitate that is denser and more easily dewatered than Mn precipitates from a physical/chemical process.

The filter must support fully aerobic conditions (i.e., significant amounts of oxygen must be present). This condition demands much more vigorous aeration than for Fe removal. Typical conditions needed within the filter are pH above 7.5, dissolved oxygen (DO) greater than 5 mg/L, and redox potential 300 to 400 mV.

The raw water must not contain materials toxic to the bacteria. The list of toxic substances and compounds is similar to that for biological Fe removal. In particular, chlorine (Cl_2) in the raw water or the backwash water is fatal to the process.

Aeration is required upstream of the filter to ensure that a dissolved oxygen (DO) level greater than 5 mg/L within the biofilter. On-line air injection, spray aeration, or cascade aeration towers are common aeration methods. High filtration rates of 10–40 m/h (4–16 gpm/ft^2) are possible for this method. A filter sand medium should have an effective size of 0.95–1.35 mm. If necessary, Mn removal can be enhanced by using a filter medium precoated with manganese dioxide, $MnO_2(s)$. Manganese greensand is too fine, having an effective size of 0.30–0.35 mm.

A new biological Mn removal plant requires a longer startup than is needed for biological Fe removal. Development of sufficient numbers of bacteria organisms may take 2–8 weeks.

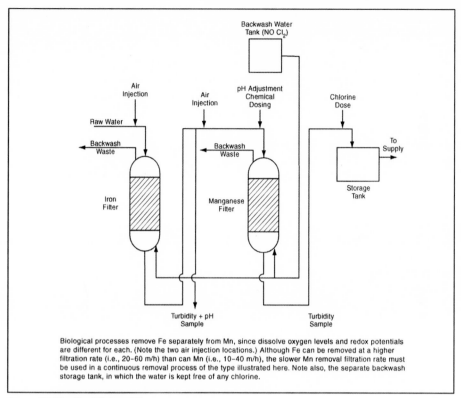

Biological processes remove Fe separately from Mn, since dissolve oxygen levels and redox potentials are different for each. (Note the two air injection locations.) Although Fe can be removed at a higher filtration rate (i.e., 20–60 m/h) than can Mn (i.e., 10–40 m/h), the slower Mn removal filtration rate must be used in a continuous removal process of the type illustrated here. Note also, the separate backwash storage tank, in which the water is kept free of any chlorine.

Source: Reid Crowther & Partners Ltd.

Figure 7-4 Biological removal of Fe and Mn

Biological Removal of Both Fe and Mn

Removing both Fe and Mn requires a two-stage process, as shown in Figure 7-4. The treatment process would include initial aeration and filtration for biological Fe removal, secondary aeration to elevate the dissolved oxygen levels again, pH adjustment above 7.5 (using lime, soda ash, or caustic soda), secondary filtration for biological Mn removal, and finally some means of disinfection (such as chlorination).

Biological Mn removal is not practiced as frequently as biological Fe removal. Where both Fe and Mn appear in the raw water, biological Fe removal is sometimes followed by chemical treatment to remove Mn by oxidation or adsorption. Published sources offer additional information about biological Fe and Mn removal (e.g., Mouchet 1992).

In-Situ Removal of Iron and Manganese

In-situ removal (see the glossary) involves the removal of Fe and Mn in the ground surrounding a production well. The process is relatively new, with only a few operating installations in Europe and the United States, but it shows a potential to offer a simple, cost-effective approach to the problem of Fe and Mn in potable water.

Several available commercial processes all use the same basic approach, as shown in Figures 7-5 and 7-6. An oxidant, usually atmospheric oxygen (which comprises 20 percent of the earth's atmosphere) is dissolved into a flow of recharge water, which is then injected into the ground surrounding a production well either by pumping back through the production well or through adjacent recharge wells. The oxidant-rich recharge water causes Fe and Mn to form hydrous oxide surfaces on surrounding soil grains, creating a treatment zone. The recharge is then terminated, and groundwater rich in Fe and Mn is drawn through this zone as it is pumped from the production well. The hydrous oxide surfaces adsorb ferrous and manganous ions, reducing dissolved Fe and Mn concentrations. When the treatment zone becomes exhausted (i.e., loses its capacity to adsorb Fe and Mn), it is reactivated by the injection of oxidized recharge water. This recharge/withdrawal cycle is repeated as needed. The duration of each cycle is determined by the volume of water that can be drawn from the production well before Fe and Mn reach unacceptable levels.

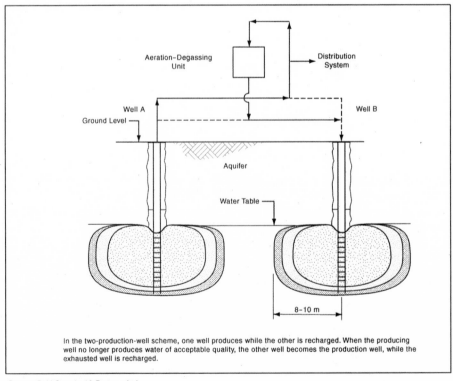

In the two-production-well scheme, one well produces while the other is recharged. When the producing well no longer produces water of acceptable quality, the other well becomes the production well, while the exhausted well is recharged.

Source: Reid Crowther & Partners Ltd.

Figure 7-5 One type of in-situ treatment system

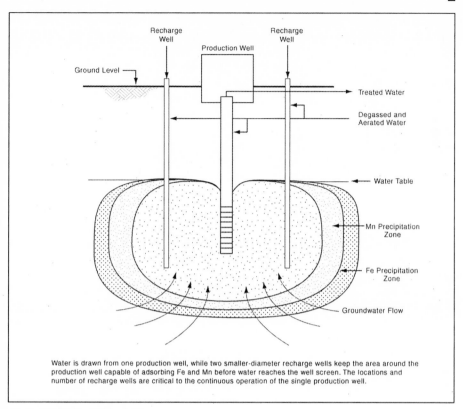

Water is drawn from one production well, while two smaller-diameter recharge wells keep the area around the production well capable of adsorbing Fe and Mn before water reaches the well screen. The locations and number of recharge wells are critical to the continuous operation of the single production well.

Source: Reid Crowther & Partners Ltd.

Figure 7-6 Recharge wells for in-situ treatment

The locations of recharge wells are most important to the success of this process. Placement must take into account flow patterns within the aquifer, the porosity of the aquifer (i.e., how easily water travels through the rock, gravel, or sand formation), and the need to draw all production water through the treatment zones established around the recharge wells.

Some systems use several wells which alternate between recharge and production modes of operation. This configuration permits the recharge and production operations to continue at the same time, providing a constant flow of treated water. Such a system, using two wells, is shown in Figure 7-6. A small proportion of the production flow from the first well is diverted through an aeration unit to the second recharge well for reinjection and restoration of the treatment zones surrounding that well. When the treatment zone surrounding the production well becomes exhausted, the roles of the two wells are reversed, and the first well becomes the recharge well.

Atmospheric oxygen can be injected in a number of ways, including forced air injection (i.e., using an air compressor). The most simple and reliable method is vacuum injection, in which water rushing past a designed suction port sucks air into the water line. Such a device is typically located next to the recycle pump. Other names used for the device are hydrocharger and air injector.

At startup, the recharge flow usually consists of untreated groundwater. However, as the treatment zone becomes fully established, a supply of treated groundwater, with reduced concentrations of Fe and Mn, is stored for use. A small system can draw recharge water back from distribution system reservoirs. The ratio of production volume to recharge volume varies according to local circumstances, but generally recharge volume can range from 10 to 30 percent of production volume. Since Mn takes much longer than Fe to oxidize, soluble Fe levels decrease rapidly following process startup, while Mn removal takes much longer to achieve acceptable levels.

A pilot study is normally required to prove suitability of the process and determine appropriate design and operating parameters. Factors affecting the process include temperature, alkalinity, the time water takes to travel through the treatment zone, and the presence of oxygen-depleting compounds within the treatment zone.

Water temperatures below 5°C slow down the process. Generally, lower alkalinity equates to lower treatment efficiency, for reasons that are not fully understood. The water being pumped for consumers should take 24 to 48 hours to travel through the treatment zone. This time varies widely from one well to the next, because the water's travel time depends on the makeup of the aquifer. Aquifers are made up of varying combinations of sand, gravel, and rock, along with other materials (including soils, coals, etc.). In this method, aquifer makeup largely controls detention time (i.e., the contact time of the water with the hydrous oxide surfaces created by the injection of oxygen-rich water back into the aquifer), just as the size of filter medium particles in a pressure or gravity filter influences the detention time there. Mineral sulfides, such as iron pyrites, or high levels of ammonia chemically inhibit the adsorption removal process.

The effective life of an in-situ treatment system depends on the well's production rate, the concentrations of Fe and Mn in the groundwater, and the porosity of the aquifer. Theoretically, the Fe and Mn hydrous oxides should gradually clog the pores of the aquifer, causing a progressive decrease in performance. European experience showed little evidence of this effect in the decade prior to 1995, however. A more significant practical problem is frequent clogging of the recharge wells, particularly during startup when recharge water quality can be poor (i.e., high in Fe and Mn, as well as other elements and compounds). Operating costs and procedures should allow for rehabilitation of recharge wells using such methods as acid or air injection twice or more a year.

Another operating problem occasionally experienced is corrosion of the well screen and pumping system. This process is accelerated by high levels of free and dissolved oxygen in the recharge water.

Operating costs are modest for in-situ treatment. The method requires no chemicals, produces no sludge to handle, and requires only intermittent supervision

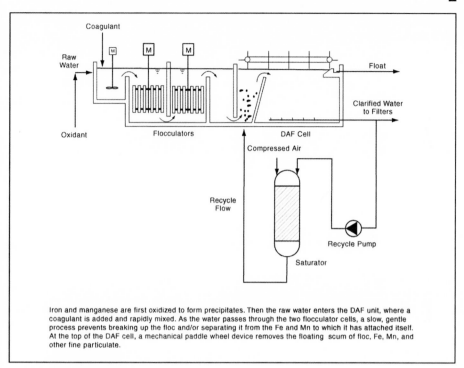

Iron and manganese are first oxidized to form precipitates. Then the raw water enters the DAF unit, where a coagulant is added and rapidly mixed. As the water passes through the two flocculator cells, a slow, gentle process prevents breaking up the floc and/or separating it from the Fe and Mn to which it has attached itself. At the top of the DAF cell, a mechanical paddle wheel device removes the floating scum of floc, Fe, Mn, and other fine particulate.

Source: Reid Crowther & Partners Ltd.

Figure 7-7 Dissolved air flotation

by operating staff. Despite the absence of chemical and sludge handling costs, however, in-situ treatment systems are unlikely to be cheaper than alternative processes at production rates less than about 1,000 L/min (264 gpm) due to the cost of establishing the recharge wells and associated storage and pumping system costs.

Dissolved Air Flotation

Dissolved air flotation (DAF) is not in itself a process for removing Fe. However, like sedimentation, it does remove the precipitates formed following the oxidation of the dissolved Fe and Mn by agents such as aeration, Cl_2, $KMnO_4$, and O_3.

A typical DAF process is shown in Figure 7-7. The Fe and Mn are first oxidized, and sufficient time is allowed for completion of the oxidation reaction. Next, a coagulant such as alum is added to the water, causing the Fe and Mn precipitates produced by oxidation to become suspended in solution, where they can clump together as flocs. Gentle stirring promotes flocculation, and the flocs grow over time to sizes that allow easy removal.

As the flocculated water enters the DAF cell, microscopic air bubbles are injected into the flow, where they attach to the particles of floc. The floc then rides to the surface on the air bubbles, where it forms a thick scum. At regular intervals, the scum is removed using rotating paddles. Clear water then leaves from the lower outlet end of the DAF cell.

Efficiency of the process demands extremely small air bubbles. To achieve this condition, clarified water from the DAF cell is pumped into a saturator, a device for mixing air and water under pressure of approximately 380–480 kPa (roughly 55–70 psi). A flow of pressurized, air-saturated water flows from the bottom of the saturator to injection nozzles at the DAF cell inlet. As the flow is forced through the nozzles, the sudden pressure change causes the air to come out of solution in the form of very fine bubbles, similar to what occurs when the top is removed from a bottle of soda or beer. Shaking the bottle first creates massive bubble release and surface froth.

The process is rapid, much quicker than sedimentation, so DAF cells are substantially smaller than settling tanks of equal flow capacity. However, DAF works best with low-turbidity waters that contain some organic material such as algae. DAF is not usually the best process to choose for treating river water with seasonally high silt loads.

Experience with DAF has shown that the process easily removes Fe precipitates, especially organically bound precipitates. While DAF will remove low levels of Mn precipitates, it sometimes encounters difficulty removing high Mn concentrations, as the precipitates aren't always easy to float. In these cases, a process might first remove Fe using a low oxidant dose, followed by DAF. Treatment would continue by removing Mn by oxidation/filtration, oxidation/coagulation/filtration, or adsorption using $MnO_2(s)$-coated filter media (such as manganese greensand or pyrolusite mixed with sand).

Caution

All technologies covered in this handbook are considered legitimate, proven methods of removing Fe and Mn. Some are more broadly used than others, and still others are benefiting from advancing knowledge.

From time to time, however, removal processes are offered that seem too good to be true. Clearly, Fe and Mn removal is not a simple task. If a system or process claims to always remove Fe and Mn easily and completely, some healthy skepticism may be warranted. Special caution should be taken when vendors try to keep secret their filter media specifications, chemicals used, or methods of removal.

All other things being equal, buyers should review some important questions in choosing a removal process:

- Does the process remove Fe below 0.3 mg/L and Mn below 0.05 mg/L?
- How does the cost compare with tried-and-true methods already available to the buyer?
- Can the system or process be sustained easily, so that it continues to work for many years (up to 15 to 20 years depending on raw-water quality)?

Pretreatment, filtration, and *sustainability* are three equal keys to successful removal of Fe and Mn. Don't buy before pilot testing the offered system or process in a specific water treatment plant. Get a list of other communities using the offered system or process, and call them to get answers to a written list of questions. Call your engineering consultant or someone you know is knowledgeable and reputable, and ask for advice.

Caveat emptor—buyer beware!

Chapter 8
Pilot Studies and Examples

∎

Purpose and Need

Water treatment design and operating professionals recognize the importance of certain critical details and assumptions in evaluating performance data. The design and execution of a performance evaluation for a treatment plant require effective planning and technical skills. Only in this way can a study produce critical information that supports goals and does not mislead.

The basic purpose of a pilot study is to establish a process for removing Fe and Mn below target levels. However, pilot studies go well beyond this basic purpose to perform other functions, including:

1. Determining a site-specific chemical treatment program. The study usually provides data for calculating yearly volumes of chemical required and estimating costs.
2. Specifying hardware items (e.g., chemical pumps, valves, instrumentation, air blowers, and underdrains)
3. Specifying filter media
4. Developing data for calculating theoretical filter runs
5. Spelling out backwash and air scour rates and procedures
6. Studies undertaken before initiating plant design provide consultants with data about appropriate filtration rates, required detention times, aeration requirements if any, sequence of chemical addition, and other prime design data.

Pilot studies help to conserve both water and money. Data developed in dozens of pilot studies indicate that most particulate removal occurs during the first 2 to 4 min of a hydraulic backwash. Many plants waste untold gallons of treated water in unnecessarily long backwashes. The same pilot studies have proven the value of air scour in loosening filtered particulate from the filter media, to be carried to the waste line with low flows of backwash water. Optimized backwash procedures like these save both water and money.

111

A pilot study conserves money in obvious ways, as shown by many examples of plants built or rehabilitated without this precaution. Design deficiencies resulting from the absence of pilot testing data include inadequate detention time, unacceptably low backwash rates, filtration rates too high to maintain, selection of filter underdrain designs based on lowest cost rather than hydraulic performance, standalone air scour grids prone to plugging, chemical pretreatment processes based on incomplete data, and many others.

Before an actual pilot study begins, a study of the raw water must be undertaken. Waters with very similar characteristics can exhibit distinctly different treatment characteristics.

Taking Water Samples

Samples for analysis in a pilot study should be collected carefully to make sure the most representative sample possible is obtained. In general, they should be taken near the center of a vessel or duct and below the surface. Use only clean containers (bottles or beakers) for collecting samples. Since nitric acid is often used to clean glassware, an operator should rinse a container several times with the water to be tested before taking a sample.

Any sample should be collected as close as possible to the source of the supply to minimize the effects of distribution system conditions. The water should be allowed to run for sufficient time to flush the system, and the sample container should be filled slowly with a gentle stream to avoid turbulence and air bubbles. A water sample from a well should be collected after the pump has run long enough to deliver water representative of the groundwater feeding the well.

It is difficult to obtain a truly representative surface water sample. Best results can be obtained by running a series of tests with samples taken from several locations and depths at different times. Results then establish a pattern applicable to the overall body of water.

Samples intended to measure soluble Fe in well water should ideally be taken without allowing contact with oxygen in the air (a difficult task) but certainly not from a turbulent stream or bubbling water. Analysis should be completed as soon as possible after the sample is taken to minimize contact time with oxygen. In some cases, Fe is oxidized almost instantly by contact with oxygen. If the sample will be transported and analysis delayed, take care to let the well stream gently flow into the sample container, fill the container to overflowing, and seal with a tight cap. This collection method keeps oxidation to a minimum. For this and other reasons, operators should analyze samples as soon as possible after collection.

Depending on the nature of testing to be conducted, other special precautions in handling samples also may be necessary to prevent natural interferences such as organic growth or loss or gain of dissolved gases. For example, samples to be sent to an independent lab for organic carbon and ammonia nitrogen analysis are preserved with sulfuric acid. Also, samples for hydrogen sulfide analysis are preserved with sodium hydroxide (NaOH) and ascorbic acid.

A good source of information on collecting and analyzing samples is the book, *Standard Methods for the Examination of Water and Wastewater* (APHA, AWWA, and WEF, 1998).*

Historical Data Requirements

In many jurisdictions, municipal water is analyzed periodically for chemicals, elements, and biocides related to health and toxicity. These analyses should be carefully reviewed for any abnormal levels of chemicals and elements that may affect the treatment process about to undergo pilot testing. Tests should also check for interference with certain analytical procedures. For example, the Hach pan method for determining low-range Mn concentrations (i.e., up to 0.7 mg/L) can be disrupted by certain levels of aluminum, cadmium, calcium, cobalt, copper, iron, lead, magnesium, nickel, and/or zinc.

General chemical analysis data often include much of the information necessary to calculate a Langelier saturation index and aggressive index. The LSI, a measure of a solution's ability to dissolve or deposit calcium carbonate scale, is often used as an indicator of water's corrosivity. The index is not related directly to corrosion, but rather to the deposition of a calcium carbonate film or scale. When no protective scale is formed, water is considered to be aggressive, and corrosion can occur. See chapter 2 for additional information.

A pilot study reviews the LSI to evaluate the need for measures to prevent calcium carbonate from covering media particles, which could change their effective size over time and/or blind-off MnO_2 sites, preventing adsorption of Fe and Mn.

Values needed to do an LSI calculation may be available from a general chemical analysis. They include total dissolved solids, calcium hardness, and total alkalinity. In addition, water temperature and pH values are required, but these data are obtained at the time of the pilot study.

A general chemical analysis may also include levels of nitrates, sulphates, and sodium. Sulphates may be important, for example, because they can cause taste problems.

Chemicals Data Developed On Site

Values for Fe and Mn in the raw water are developed on site at the time of the pilot study. Color, temperature, and pH should likewise be determined. Because no method gives reliable forecasts of occurrences within an aquifer, these five values (Fe, Mn, pH, color, and temperature) should be measured several times during the course of a pilot study. As some actual pilot study data given later in the chapter demonstrate, many measurements of Fe and Mn may be made during a filter run.

---

*Available from American Water Works Association, 6666 W. Quincy Ave., Denver, CO 80235, (800)926-7337.

These levels may fluctuate, with dramatic variations in some well waters. In such a case, the treatment process selected based on pilot study data must compensate for fluctuations in raw-water chemistry.

Data Collected During a Pilot Study

1. **Influent and effluent quality**
   The first results of a pilot study indicate initial removal efficiency. One of the main functions of a pilot study is to indicate both which processes work and which do not.

2. **Filtration performance as a function of depth**
   A pilot study looks for very important information about where the filtration takes place in a filter. If much of the filtration takes place in the top regions of the filter bed, perhaps the design has specified too fine a filter medium. An adsorption process should ideally keep Fe out of the $MnO_2(s)$ layer to prevent coating of the particles with iron floc and development of head losses that could reduce filter run lengths. To evaluate this possibility, data about Fe and Mn penetration must be developed based on time and depth.

3. **Head loss during a filter run**
   For either an open-top or pressure filter, a pilot study must measure head loss during filter runs. For an open-top gravity filter, this information indicates the length of the filtration cycle based on available driving head (i.e., the capacity to push water through a filter without overflowing the filter compartment). For a pressure filter using manganese greensand, a study must determine the length of time needed to accumulate a pressure differential of 55 kPa (8 psi). Pressure differentials in excess of this level can result in fracturing of greensand grains and loss of filter performance over time. For pressure filters not using manganese greensand, a study must determine the length of time required for the filter bed to reach its solids-holding capacity; at this point channeling commences, indicated by Fe or Mn breakthrough (i.e., rapidly rising residual values).

4. **Filter ripening time**
   Filter ripening is defined as the improvement in water quality from the start of a filter run until optimum quality is achieved (whatever parameters determine this quality level in a specific water treatment plant). Ripening time varies from filter to filter depending on many factors, such as nature of the influent raw water, filter flow rate, filter media design, chemical pretreatment program, etc. To achieve the objective of distributing only water of optimum quality, filtered water should run to waste from the beginning of a filtration cycle until optimum quality filter effluent results. If a filter is designed to start-stop-start a number of times between backwash cycles based on clear well levels, the length of filter ripening each time can be established in a pilot study.

5. **Filter media design**

   Several filter media designs can be tested if required, to prove or disprove performance expectations. A media design that performs well in one plant may not be suitable for another, despite similarities in raw waters and chemical pretreatment programs. Pilot testing of media designs also provides an opportunity to evaluate preferences.

6. **Filter loading rate**

   This term describes the maximum filtration flow rate at which optimum quality water can be produced, holding constant all other variables. Evaluating this rate avoids problems that could occur despite a correct chemical pretreatment program, appropriate piping configuration in the plant, and good filter media bed, because the rate of flow through the filter (i.e., loading rate) is too high to produce water of the quality desired. More likely, loading rate data reveal a need to design a new media bed to increase the plant's daily capacity or produce better quality water at the current filter loading rate. If production volumes are flexible, the pilot study may indicate at what reduced filter loading rate the existing plant could produce optimum water quality.

7. **Establishing backwash procedures**

   The right design for the removal process (i.e., chemical pretreatment, media bed, filtration rate, etc.) is of little value if the process cannot be sustained. Deteriorating water quality, declining filter run lengths, and increasing need for backwash water may indicate failure of the underdrain to distribute backwash water evenly, leaving unacceptably high levels of filtered particulate in parts of the media bed. These problems may also result if backwash flow rates are either too low or too high, and/or the wrong backwash procedure is used. At the end of each pilot filter run, empirical data are developed. Accumulating data guide development of a procedure to optimize bed cleaning using the minimum amount of time, backwash water, and air scour.

8. **Impact of the chemical program on filter media**

   Pilot testing with filter columns made of clear acrylic allows operators to see filter media agglomeration. This important observation is very valuable in establishing an effective media bed cleaning procedure. The behavior attributed to pretreatment chemicals could prompt rethinking of a proposed media bed design. For example, the use of a certain polymer may require a deeper or coarser coal layer. Conversely, testing of a particular media combination may indicate particle agglomeration at the interface of the coal and the layer below, indicating the top coal layer is too coarse and/or too shallow for effective filtering.

9. **Optimizing pretreatment chemicals**

   Jar testing is the most common method used to arrive at an optimum chemical dosage. While such a test can indicate whether the chemical chosen will form a suitable floc, it cannot predict the filterability of the developed floc. Visual observation of a chemical reaction of this nature can too easily be influenced by personal biases, which lead to inappropriate conclusions concerning optimum feed rates. Instead, measurement of both dosages and effluent quality can lead toward an optimal pretreatment program.

Empirical Data: Examples from Actual Plants

Water treatment is a site-specific process. Methods that remove Fe and Mn from one well water may not remove them from another pumped just miles down the road. Pilot testing may be required to find the right chemical coagulant, for example, because one that works well in one location may not perform as well in another. Variations in the levels of Fe and Mn in the raw water require variations in the depth of filter media in a dual-media bed. Chemical demands also need reevaluation from time to time.

The early part of this chapter has covered the operating parameters established in pilot studies. The rest of the chapter reviews a few samples of actual information developed in water treatment plants. Much of the information was generated in municipal water treatment plants and some in industrial plants. Some industrial plants, such as pulp-and-paper plants, require treated water of higher quality than a municipal plant produces, while other commercial users often do not.

Regardless of the water quality required, pilot testing gives a much easier, faster, and more cost-effective way to develop wanted information than a similar study in a full-scale water treatment plant of any size. This section provides an example of some types of information generated during an actual pilot study and filter audit. These brief reviews point out the need to develop information in an organized, manageable way. Bear in mind, however, that the information applies only to the location in which it was generated and should not be applied in any other plant.

Example 1: A Town's Treatment Plant

In this portion of the pilot study, the technician was searching for the appropriate chemical feed level. Having just established an approximation, an actual filter run was commenced in which the chemical feed was increased and decreased, and both turbidity and color were measured. In this filter run, a chemical dosage of 24.57 mg/L produced the best results. Table 8-1 presents an excerpt from the study report.

Example 2: Another Town

As indicated in this portion of a pilot study, results are not always as anticipated. Further testing revealed the true chlorine demand. Below is an actual excerpt from the study report.

Chlorine demand tests from pilot filter run 1

During pilot filter run #1, the chlorine dosage requirements were evaluated to determine the chlorine demand for both treatment and

Table 8-1 Comparison of turbidity and color removal efficiencies from pilot filter run 4

Run Time (Hours)	Turbidity (NTU)		Color (Units)		Chemical Feed (mg/L)
	Raw Water	Pilot Filter	Raw Water	Pilot Filter	
1	1.99	0.19	30	4	24.57
2	1.95	0.20	31	4	25.03
3	1.89	0.19	31	5	25.03
4	2.00	0.21	30	4	24.11
5	1.87	0.21	31	4	25.03
6	1.91	0.22	31	4	25.03
7	1.93	0.23	30	5	24.57
8	1.96	0.23	30	7	25.03

Enhanced coagulation with a polyaluminum chloride coagulant. AWI mono media 0.65-mm crushed quartz sand, 36 in. depth. Service flow rate 4.33 gpm/ft^2.

disinfection with the Pyrolox matrix media in the continuous regeneration mode. A concentrated sodium hypochlorite (NaOCl) solution was prepared containing 1,380 mg/L of available chlorine. The actual chlorine feed rate was then determined by direct measurement of the concentrated NaOCl solution with a graduated cylinder to the volume of water treated. From the chlorine test results, breakpoint chlorination was achieved when the chlorine dosage was in the range of 9–10 mg/L. This was above the expected chlorine demand for ammonia-nitrogen and iron and manganese oxidation which would have been in the range of 5 mg/L. The excess chlorine demand was predominately due to the installation of new filter coal media with the Pyrolox matrix media in the pilot filter unit. New filter coal media, typically, has a temporary chlorine demand and conditioning of the media bed with a concentrated chlorine solution should be performed prior to the initial filter operation. Chlorine dosage requirements from pilot filter run #1 are tabulated in [Table 8-2].

Example 3: Commercial WTP

The chemical pretreatment program in this plant and the operating procedures indicated a choice of a filter bed with an optimum solids-holding capacity. How would that capacity compare to the capacity of a highly angular sand of the same size as the round sand currently in use? Also, does an increase in solids-holding capacity translate to an increase in backwash water? These same questions apply when designing a filter bed of coal over manganese greensand.

Side-by-side, portable 15 cm (6 in.) diameter filter columns were fed from the same pretreated water source and both filtered at the same rate for the same length

Table 8-2 Chlorine dosage requirements from pilot filter run 1

Run Time (hours)	Initial Reading (mls)	Final Reading (mls)	Cl_2 Solution (mls)	Cl_2 Feed (mg/L)	Pilot Filter Effluent	
					Free Cl (mg/L)	Total Cl (mg/L)
0.5	1,990	1,890	100	6.76	0.08	0.81
1.0	1,890	1,755	135	9.13	0.31	1.16
1.5	1,755	1,595	160	10.82	1.20	1.42
2.0	1,595	1,425	170	11.49	1.33	1.63
2.5	1,425	1,240	185	12.51	1.47	1.75
3.0	1,240	960	280	18.93	2.84	3.29

of time. Then both were identically backwashed. Turbidity levels indicated that the angular sand did accumulate more particulate, as expected. High levels of bed cleaning were achieved with identical backwash volumes, as indicated in Table 8-3, an actual excerpt from the study report.

Example 4: A Third Town

In this 12-hour run, the technician wanted to know levels of Fe and Mn removal, as well as levels of Fe and Mn in the raw water. An excerpt from the study report follows

In pilot filter run #2, the Pyrolox matrix media continued to be operated in the direct filtration mode with a sodium hypochlorite (NaOCl) prefeed as a continuous regenerant. The filter run was increased to 12 hours to

Table 8-3 Backwash test results-Angular sand versus silica rounded sand

Backwash Time (min)	Turbidity (NTU)		Comments
	0.5-mm Angular Sand	Silica Sand	
0–6			Air scour (3.5 scfm/ft^2)
0.5	457.0	248.0	Hydraulic backwash (20 gpm/ft^2)
1.0	248.0	95.6	
2.0	41.5	20.2	
3.0	28.7	9.8	
4.0	10.6	5.8	
5.0	10.6	5.8	
6.0	3.7	1.9	
7.0	2.4	1.3	
8.0	1.5	1.2	End hydraulic backwash

evaluate the performance of the Pyrolox process during an extended filter run. The Pyrolox matrix media configuration remained unchanged from the previous pilot filter run and the service flow rate was maintained at 2.1 gpm/ft^2. The removal efficiencies during the course of pilot filter run #2 ranged from 98.3 percent–99.0 percent for iron (Fe) and 94.8 percent–95.3 percent for manganese (Mn), respectively. In the extended filter run, both iron and manganese were removed to within the provincial drinking water quality objectives (DWQO) of < 0.3 mg/L and < 0.05 mg/L for iron and manganese, respectively. Test results from pilot filter run #2 are tabulated in [Table 8-4].

Table 8-4 Test results from pilot filter run 2

Run Time (hours)	Raw Water		Pilot Filter	
	Fe (mg/L)	Mn (mg/L)	Fe (mg/L)	Mn (mg/L)
1	1.99	0.498	0.03	0.026
2			0.03	0.024
3	1.81	0.477	0.03	0.024
4			0.03	0.024
5			0.02	0.025
6	1.86	0.481	0.03	0.024
7			0.02	0.025
9	1.85	0.476	0.02	0.024
12	2.02	0.487	0.02	0.023

Chapter 9
Distribution System and Safety Issues

<hr>

Distribution System Cleaning

Regardless of how capable a water treatment plant's filters are or how efficiently the operator has removed Fe and Mn below objective levels, some fine silts and particulate oxidation by-products settle out in low-flow areas of the distribution system. Simply opening a fire hydrant may reverse flows in a portion of the distribution system and fluidize accumulated silts. Consumers in the vicinity are likely to get periods of brown water, the term applied to any off-colored water, regardless of the cause of discoloration.

As a rule of thumb, distribution lines should be swabbed about every 3 to 5 years, even if the lines have carried only top-quality water. In addition to swabbing, annual maintenance should include flushing at least once a year. (Most communities flush in both spring and fall.) Obviously, high levels of Fe and Mn in treated water call for more frequent distribution system cleaning.

Uncleaned lines are also ideal locations for problems with iron bacteria. These organisms colonize and grow, then slough off chunks which end up as brown water in the homes and businesses of consumers. (See Chapter 4, Microbiology of Iron and Manganese Removal.)

Squeaky clean lines deliver water without taste or odor problems. These conditions also provide a poor breeding ground for any organisms that may be harmful to humans.

Who Should Do the Work?

If a public works department has adequate staff and the available budget to buy necessary equipment, it can undertake distribution line swabbing without assistance from cleaning contractors. In many small communities, however, both staff and budget are worked to the limit. Their best resource allocations often call for hiring outside cleaning contractors.

Many large communities with available resources have considered cost-effectiveness and chosen to hire outside contractors who specialize in water and sewer line cleaning. A reputable cleaning contractor often does a far more thorough job in much less time and at lower cost than an inexperienced crew learning as they go. Probably every cleaning contractor owns specially made tools, unavailable through supply companies, that make the job easier and faster to complete.

Doing it yourself. Operators determined to proceed on their own should first develop a detailed plan for the job, specifying who does what where. This planning process includes getting firsthand information from neighboring communities, if the operators have not done the job. A careful estimate of how long the job will take is an important part of the plan. Several days before the swabbing commences, advise all affected consumers how long service will be interrupted. This notice can be communicated by telephone, newspaper, public-service announcements on radio and TV, community TV channels, face-to-face/door-to-door, mailed flyers/letters, or a combination. This critical part of the job requires careful handling. Remind consumers just before you start work how long they will go without water, and keep them informed if delays take place. **Handling public relations right is just as important as getting the lines clean.**

Advise consumers to flush their cold water lines from outside taps after the swabbing job is done to prevent dirty water from reaching dishwashers, water softeners, domestic water heaters, or clothes washers. This protection is especially important for medical equipment, sensitive commercial equipment, etc.

Line cleaning and hydrant flushing are fair-weather jobs. Obviously, doing this work in freezing temperatures risks coating equipment with ice. Water leaving the hydrant begins to freeze in place, and workers must contend with slippery surfaces and wet clothing and gloves. Also, thorough clean-up is an impossibility.

In warm seasons, however, flushing treated water through a hydrant may cause problems during periods of peak demand when filtration units and/or storage reservoirs are barely keeping up with requirements. During swabbing operations, always assure an adequate supply of water for fire-fighting emergencies.

Many communities flush hydrants to clean distribution lines. Whether or not flushing is an adequate cleaning measure depends on the quality of the water flowing through the distribution lines and the frequency of flushing.

The best combination of practices is to swab once every 3 to 5 years and flush hydrants annually or twice a year to discharge loose, unwanted material in the lines and exercise the hydrants' internal mechanisms. The needed frequency of swabbing can be estimated based on water quality in the distribution system.

Because the water leaving the water treatment plant has a chlorine residual, some oxidation of untreated iron, manganese, and organic material takes place in the distribution lines. These products of oxidation should be systematically removed. During repairs of water main breaks, workers should check conditions inside pipes. Obviously, if lines remain clean, swabbing would be a waste of money.

Procedures for Cleaning Distribution Lines

Swabbing involves passing a foam plug through the distribution lines. Workers insert a compressible plug larger in diameter than the lines to be cleaned into a hydrant or other point designed for that purpose. They force it down the insertion line or hydrant using pressure from a fire department pumper truck or a portable pressure pump, then distribution line pressure carries the plug along to a hydrant which has had its internal mechanism removed so the plug can exit. This sequence seems simple, but complications often arise.

Small communities often buy too few foam plugs, so they must delay work while more are delivered. A better practice is to buy many more than the estimated number needed and return the unused plugs for a refund. A typical foam plug has a coarse material glued to the surface to act as sandpaper does; an extensive variety of designs and coarseness ratings suit specific applications, but anyone not in the cleaning business may not know which to buy.

As a part of the planning stage, valves and hydrants should be exercised to confirm they will work without difficulty once the swabbing commences. Faults should be repaired before starting the swabbing program. A small inventory of spare parts should be on hand, in case breakdowns occur during the swabbing operation. Be sure all the hand and power tools needed for the swabbing job are ready to go. Before the first time cleaning lines, a crew will benefit from a practice run. An hour or two going through the actual motions without shutting down lines or inserting swabs or plugs will pay big dividends when actual work starts.

Swabs or plugs may become lodged in lines and go no further. A stuck swab will break apart over time if left in place, and small chunks could then lodge in valves and water meters or in specialized equipment. Removal of a stuck plug is normally accomplished by reversing the flow of water.

Systematic cleaning in the shortest time takes good two-way radio or telephone communication from the point of insertion to the point of exit. Small communities may not have the communications capability, which extends the time required to get the job done.

Both swabbing and hydrant flushing should be done systematically starting nearest the water treatment plant and working away from it in an ever-increasing radius until line ends are reached. Small amounts of silt, etc. can simply be flushed down the street into the nearest drain. However, any amount over a shovelful should be captured and removed to a dump or fill site. If a swab is needed to move Fe and Mn build-up, silt, etc. out of a pressure line, it will likely settle into a new home in a gravity sewer line if flushed there. Sooner or later, the accumulation will create a second cleaning job.

Care should be taken to direct the flows from an open hydrant to avoid gushing onto private property or interfering with street traffic patterns. A series of plywood baffles can be erected to control flow direction and trap silts, etc. Adequate labor should be available to keep the swabbing in progress while a clean-up crew follows behind to restore previous conditions.

Benefits of Distribution Line Cleaning

1. Potential health risks from waterborne organisms are reduced.
2. Clear water, free of taste and odor picked up in the distribution system, is delivered to consumers.
3. Brown water complaints due to accumulated deposits in the distribution system are almost eliminated.
4. Chlorine residuals are more easily maintained, and chlorine costs are reduced, because material removed by swabbing would have a chlorine demand.
5. Hydrants and street valves are operated, creating an opportunity to record their performance and organize maintenance work where necessary.
6. The amount and cost of sequestrants (discussed in chapter 7) will be reduced, because cleaning reduces lines' demand for the chemicals.

Safety Issues

Handling Chemicals Safely

Treatment chemicals and reagents for chemical tests typically pose potential threats to an operator's skin, eyes, nose (mucous membranes), and/or lungs. Some testing reagents pose greater hazards than others. For example, anyone measuring Mn by the Pan method should not let the reagents come into contact with the skin. Trembling or unsteady hands are best protected by tight-fitting latex gloves. If a medical condition causes trembling hands, a full-length polyvinyl chloride (PVC) apron and clear acrylic full-face shield should be worn during routine daily residual testing.

Almost all water treatment plant (WTP) chemicals should be handled only by workers wearing certain protective clothing and devices. Anyone handling dry chemicals such as alum, soda ash, calcium hypochlorite, and potassium permanganate crystals should wear a full-length PVC apron, elbow-length chemically nonreactive gloves, goggles or full-face shield, disposable or replaceable cartridge-filter-type dust mask, and cap. When handling large numbers of chemical bags or sacks in one session, or handling drums or cylinders, steel-toed boots should be worn.

Many small plants use sodium hypochlorite as a chlorine source, diluting the product with water before dosing. Sodium hypochlorite (NaOCl) mixed with water forms hypochlorous acid (HOCl), a severe skin irritant, so long sleeves or other protective clothing should be available. Those who are concerned about potential corrosiveness of HOCl or staining by $KMnO_4$ on foot wear should consider using rubber boots when handling these two chemicals. Dry calcium hypochlorite can cause burns if it touches the skin or eyes.

CAUTION: Every plant using chlorine gas should maintain two air packs fully charged and available for use at all times. Chlorine gas breathed in can quickly (within a few seconds) kill the breather or at least make him/her incapable of doing anything, including leaving the chlorine room. The second air pack should always be available as a backup, or for use by someone responding to the needs of a fellow worker in trouble in the chlorine room.

Activated carbon is highly reactive with many other substances and fluids. No one should breathe in powdered activated carbon dust. Consult a safety supplier for appropriate replaceable filter cartridges. Some manufacturers of filter cartridges have developed a single cartridge that covers most eventualities likely to arise in small plants.

Every WTP should collect a complete set of up-to-date material safety data sheets (MSDS) covering every treatment chemical and chemical reagent used in that plant. Every employee should know the location of the set and be familiar with the information these documents contain. Guided by the complete set of MSDS and common sense, first-aid supplies should be available and easily accessible in the plant. If such devices as free-standing showers or eye-bathing stations are not available, an adequate supply of flushing–washing water should be kept available at all times. Any WTP operator should know from memory what chemicals can be flushed with water from eyes, skin, and nose and for how long; whether or not vomiting should be induced; what type of bandaging is appropriate; under what circumstances hospital/medical clinic or ambulance services are required; and whether or not items of clothing should be removed or additional covers for warmth should be added.

A final reminder: after handling chemicals, always thoroughly wash your hands before touching your eyes, mouth, nose, or food. The eyes and mucous membranes of the nose are especially sensitive areas.

Dangers of Hydrogen Sulfide

CAUTION: Hydrogen sulfide can kill you. Under the right reducing conditions, dangerous levels of hydrogen sulfide can be generated within an enclosed filter vessel, a good reason why all pressure filters should be fitted with adequately sized air-release valves (not pressure-release valves).

The *Chemical Hazard Summary*, published by the Canadian Centre for Occupational Health and Safety (1985), describes hydrogen sulfide as a colorless gas, or a colorless liquid at low temperature or high pressure, with an offensive, strong odor similar to rotten eggs at low concentrations. The odor becomes sweetish at moderate concentrations before the sense of smell is lost. The CCOHS publication states, "The typical rotten egg odor of hydrogen sulfide can be smelled at very low concentrations. At 20 to 30 ppm the odor becomes intense. Exposure to concentrations of 100 ppm may cause rapid loss of the sense of smell. Paralysis of the nerves involved in smell may occur at 150 ppm. This means that the warning odor signal of hydrogen sulfide is lost. This loss of the sense of smell may occur gradually during exposures to small quantities of the gas or very rapidly where lethal concentrations are present."

Always use a blower to blow fresh air into a pressure vessel before entering it, and continue blowing fresh air until the work inside the filter is done. Don't rely on your nose; it may not always pick up the rotten-egg odor. Table 9-1 lists reported acute effects of exposure to hydrogen sulfide.

A number of general procedures are suggested for first aid involving poisoning by hydrogen sulfide.

Inhalation. *Persons responding to the emergency must take all precautions to ensure their own safety before attempting a rescue.* For example, appropriate respiratory protection must be worn, and the buddy system must be used. The vic-

Table 9-1 Effects of hydrogen sulfide

Concentration, *in ppm* (same as mg/L)	Effect
0.13	Odor threshold: obvious and unpleasant odor like rotten eggs, sore eyes
above 20	Intense but tolerable odor
above 50	Marked irritation of the eyes and respiratory tract (Prolonged exposure may cause bronchitis and pneumonia.)
above 100	Loss of smell, stinging of the eyes and throat (Fatal exposure can occur in 8–48 h.)
above 200	Nervous system depression, headache (Prolonged exposure may cause fluid accumulation in the lungs; fatal exposure can occur in 4–8 h.)
above 300	Life-threatening accumulation of fluid in the lungs (Fatal exposure can occur in 1–4 h.)
above 500	Excitement, headache, dizziness, and staggering followed by unconsciousness and respiratory failure within 5 min to 1 h (Fatal exposure can occur in 30 min–1 h.)
1,000	Rapidly produces unconsciousness followed by death in a few minutes

Source: Chemical Hazard Summary.

tim should be moved to an uncontaminated area. If breathing has stopped, artificial respiration or cardiopulmonary resuscitation should immediately be given by a trained person. While its effectiveness has not been clearly established, oxygen may be beneficial if administered by a person trained in its use, preferably on a physician's advice. General supportive measures (warmth, rest, reassurance, and comfort) should be provided. *Medical attention should immediately be obtained.*

Eyes. If irritation occurs due to gas exposure, the affected eye(s) should be flushed for at least 20 min with lukewarm, gently flowing water while holding apart the upper and lower eyelids. *Medical attention should immediately be obtained.*

NOTE: A physician and/or the nearest poison control center should be consulted for all cases of overexposure.

Appendix A
Calculation Tools

Table A-1 General conversion table

Multiply	By	To Obtain
Acres	0.4045	Hectares
Centimetres (cm)	0.0328	Feet
Centimetres	0.3937	Inches
Centimetres	0.0100	Metres
Centimetres	10	Millimetres
Cubic feet	1,728	Cubic inches
Cubic feet	0.0283	Cubic metres
Cubic feet	0.0370	Cubic yards
Cubic feet	7.4810	US gallons
Cubic feet	6.2280	Imperial gallons
Cubic feet	28.32	Litres
Cubic feet of water (60°F)	62.37	Pounds
Cubic feet /minute (cfm)	472.0	cubic centimetres/second
Cubic feet /minute	4.72×10^{-4}	cubic metres/second
Cubic feet /minute	0.0283	cubic metres/minute
Cubic feet /minute	0.1247	US gallons per second
Cubic feet /minute	0.4719	Litres/second
Cubic feet /minute /foot^2	5.08	Litres/second/square metre
Cubic feet /minute /foot^2	0.3048	cubic metres/minute/square metre
Cubic feet/second	0.0283	cubic metres/second
Cubic inches	16.39	Cubic centimetres
Cubic inches	0.0164	Litres
Cubic meters (m^3)	35.313	Cubic feet
Cubic meters	264.2	US gallons
Cubic meters	220.0	Imperial gallons
Cubic meters	10^3	Litres
Feet	30.48	Centimetres
Feet	0.3048	Metres
Feet of water	0.4335	Pounds per square inch

Table A-1 General conversion table—*continued*

Multiply	By	To Obtain
Feet of water	3.0	Kilopascals
Gallons, Imperial	1.2009	US gallons
Gallons, US	0.8327	Imperial gallons
Gallons, US	0.1337	Cubic feet
Gallons, Imperial	0.1606	Cubic feet
Gallons, US	3.785×10^{-3}	Cubic metres
Gallons, Imperial	4.546×10^{-3}	Cubic metres
Gallons, US	3.785	Litres
Gallons, Imperial	4.546	Litres
Gallons, US/minute/foot^2	2.4476	Metres/hour
Gallons, Imperial/minute	0.0758	Litres/second
Grains	0.0648	Grams
Grams (g)	0.0353	Ounces
Grams	2.205×10^{-3}	Pounds
Hectares	2.47	Acres
Inches	2.540	Centimetres
Inches of water	5.204	Pounds per square foot
Inches of water	0.0361	Pounds per square inch
Kilograms (kg)	2.2046	Pounds
Kilograms/square metre	1.422×10^{-3}	Pounds per square inch
Kilopascals (kPa)	0.1450	Pounds per square inch
Litres (L)	0.0353	Cubic feet
Litres	10^{-3}	Cubic metres
Litres	0.2642	US gallons
Litres	0.22	Imperial gallons
Litres/minute (L/m)	0.2642	US gallons per minute
Litres/minute	0.22	Imperial gallons per minute
Metre of water	9.8	Kilopascals
Metres (m)	3.2808	Feet
Metres	39.37	Inches
Metres	1.0936	Yards
Square metres	10.764	Square feet
Metres/minute	0.0547	Feet per second
Metres/second	3.2808	Feet per second
Microns	3.937×10^{-5}	Inches
Milligrams/litre (mg/L)	1	Parts per million (ppm)
Ounces	28.35	Grams (g)

■

Table A-1 General conversion table—*continued*

Multiply	By	To Obtain
Parts per million (ppm)	0.0584	Grains/US gallon
Parts per million	0.7016	Grains/Imperial gallon
Parts per million	8.345	Pounds/million U.S. gallons
Pounds (lb)	7,000	Grains
Pounds	453.6	Grams
Pounds per square foot (lb/ft^2)	0.0479	Kilopascals
Pounds per square inch (psi)	2.307	Feet of water
Pounds per square inch	0.0703	Kilograms/square centimetre
Pounds per square inch	703.1	Kilograms/square metre
Pounds per square inch	6.895×10^3	Pascals (pa)
Pounds per square inch	6.895	Kilopascals
Square feet (ft^2)	144	Square inches
Square feet	0.0929	Square metres
Square inches	6.45	Square centimetres
Square yards	0.836	Square metres
Square miles	2.59	Square kilometres
Temperature ($^\circ$C) + 17.8	1.8	Temperature ($^\circ$F)
Temperature ($^\circ$F) - 32	5/9	Temperature ($^\circ$C)
Yards	91.44	Centimetres
Yards	0.9144	Metres

Table A-2 Names and symbols

Name	Symbol
Centimetres	cm
Cubic centimetres	cm^3 (cc)
Cubic metres	m^3
Degrees Celsius	$^\circ$C
Grams	g
Hectares	ha
Imperial gallons per minute per square foot	Imp. gpm/ft^2
Kilograms	kg or Kg
Kilopascals	kPa
Kilowatt	kW
Litres	L
Metres	m
Micrograms per litre	μg/L

Table A-2 Names and symbols—*continued*

Name	Symbol
Milligrams	mg
Millilitres	mL
Milligrams per litre	mg/L
Millimetres	mm
Pascals (Pa = Nm^2)	Pa
Newtons	N
Square centimetres	cm^2
Square metres	m^2
US gallons per minute square foot	US gpm/ft^2

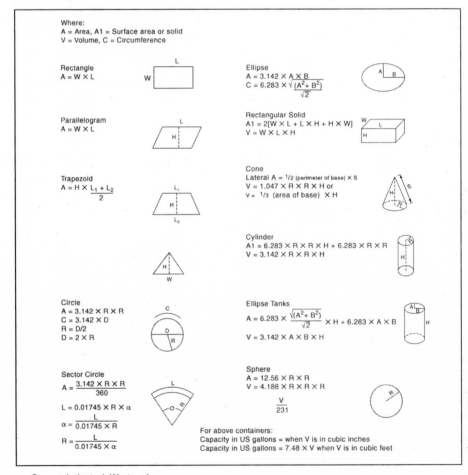

Where:
A = Area, A1 = Surface area or solid
V = Volume, C = Circumference

Rectangle
$A = W \times L$

Parallelogram
$A = W \times L$

Trapezoid
$A = H \times \dfrac{L_1 + L_2}{2}$

Circle
$A = 3.142 \times R \times R$
$C = 3.142 \times D$
$R = D/2$
$D = 2 \times R$

Sector Circle
$A = \dfrac{3.142 \times R \times R}{360}$
$L = 0.01745 \times R \times \alpha$
$\alpha = \dfrac{L}{0.01745 \times R}$
$R = \dfrac{L}{0.01745 \times \alpha}$

Ellipse
$A = 3.142 \times A \times B$
$C = 6.283 \times \dfrac{\sqrt{(A^2 + B^2)}}{\sqrt{2}}$

Rectangular Solid
$A1 = 2[W \times L + L \times H + H \times W]$
$V = W \times L \times H$

Cone
Lateral $A = 1/2$ (perimeter of base) $\times$ S
$V = 1.047 \times R \times R \times H$ or
$V = 1/3$ (area of base) $\times$ H

Cylinder
$A1 = 6.283 \times R \times R \times H + 6.283 \times R \times R$
$V = 3.142 \times R \times R \times H$

Ellipse Tanks
$A = 6.283 \times \dfrac{\sqrt{(A^2 + B^2)}}{\sqrt{2}} \times H + 6.283 \times A \times B$
$V = 3.142 \times A \times B \times H$

Sphere
$A = 12.56 \times R \times R$
$V = 4.188 \times R \times R \times R$
$\dfrac{V}{231}$

For above containers:
Capacity in US gallons = when V is in cubic inches
Capacity in US gallons = 7.48 × V when V is in cubic feet

Source: Anthratech Western Inc.

Figure A-1 Geometric formulas

Appendix B
Particle Counters in Iron and Manganese Removal

Some Fe and Mn removal processes precipitate the metals by using oxidation to form relatively insoluble substances, such as ferric hydroxide and manganese dioxide. When such a process fails to remove Fe and Mn, it is important to determine which step has failed (i.e., oxidation, precipitation, or filtration) so that the proper solution can be implemented.

Analytical methods help operators to differentiate between, for example, ferrous and ferric iron. (Consult *Standard Methods* [APHA, AWWA, and WEF 1998].) The presence of ferrous iron (Fe^{2+}) instead of ferric iron (Fe^{3+}) after oxidation indicates that this step has failed. The conditions (i.e., pH, temperature, reaction time) may be inadequate, the oxidant dosage may be insufficient, or some form of inhibition (e.g., high levels of silica in the raw water) may be causing problems.

Oxidized Fe and Mn may fail to precipitate in the presence of complexing or stabilizing agents. Examples include chelants such as EDTA, naturally occurring organic polycarboxylic acids such as the color-causing humic and fulvic acids, and silicate. These compounds form stable solutes with both ferrous and ferric iron. On the other hand, the filtration stage may fail due to inadequate operating conditions. Particles may be too fine for filter media to trap, or problems may result from lack of coagulation or flocculation, excessive filtration velocity, dirty filter media, and other causes. Analysis can determine whether the problem is with the precipitation or filtration steps by examining the particles in the filter influent and effluent.

The assessment is difficult in plants that utilize methods other than catalytic filters like manganese greensand, pyrolusite, Birm™, etc. In such a plant, oxidation and precipitation are carried out upstream of filtration, typically in a mixed tank. Other chemicals such as coagulants or flocculants may be added, and filtration is usually one of the last operations. As the treatment steps take place within different pieces of equipment, each step can be analyzed independently to determine which one is failing, leaving Fe and Mn in the water.

One of the methods used to assess whether precipitation and filtration performances is analysis of particulate in the water. This analysis can be done using either membrane filtration (explained in chapter 6) and microscopic examination, turbidity meters, or particle counters. Membrane filtration and turbidity meters cannot provide all the data needed, and the results generally depend on the nature of the particulate. Changes such as the precipitation of new particles following oxidation complicate comparisons. Also, turbidity meters may not be sensitive enough to detect small changes in particle quantities in some waters.

Types of Particle Counters

Particle counters are analytical instruments that can count the number of particles in a given volume of water. Usually, they also determine the approximate size of each particle. These very sensitive instruments are also used to monitor filter performance in water treatment plants that wish to maintain very low treated water turbidities, such as below 0.1 NTU.

An excellent reference on particle counters is the report *Evaluation of Particle Counting as a Measure of Treatment Plant Performance* (Hargesheimer et al. 1992). The following paragraphs briefly describe the techniques available based on information contained in this reference.

The three basic types of particle counter sensors are light obscuration, light scatter, and electrical resistance sensors. The first two utilize laser beams passing across capillary tubes through which samples are pumped. The third uses an electrolytic

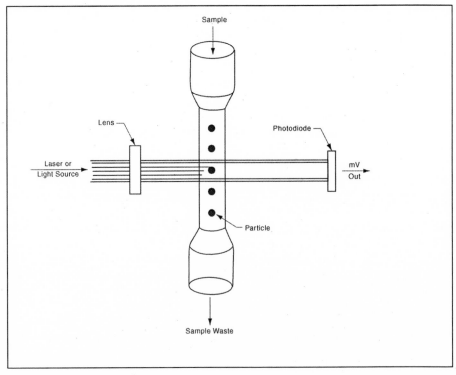

Source: Reid Crowther & Partners Ltd.

Figure B-1 Schematic of a light obscuration particle count sensor technique. All are calibrated using suspensions of tiny glass beads of known size. Particle sizes from 0.1 to 500 μm can be detected, depending on the instrument used. The techniques are adequate for dilute suspensions; turbid waters need dilution using membrane-filtered water.

Principles of Operation

In light obscuration (see Figure B-1), a particle passing through the laser beam projects a shadow on the light detector located directly in line with the beam. Each obscuration event is counted as a particle. The sensed reduction in light intensity at the detector is converted to shade area. The corresponding particle diameter is estimated from that information.

Light scatter (see Figure B-2) is based on the association between an increase in particle size and an increase in the angle of scattering of light incident on the particle. Particles passing through a laser beam scatter the light. Each scattering event is counted as a particle. Light sensors around the cell measure the degree of scatter, which is correlated with an estimated particle diameter.

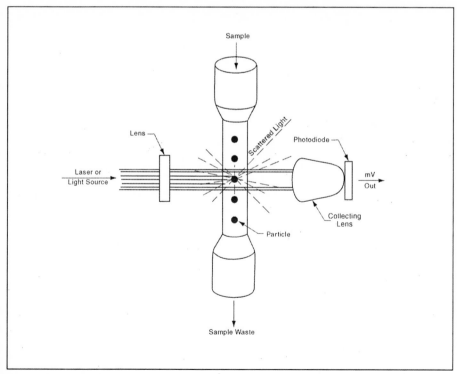

Source: Reid Crowther & Partners Ltd.

Figure B-2 Schematic of a flow-based light scatter particle count sensor

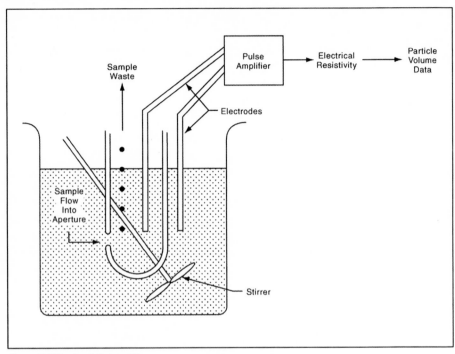

Source: Reid Crowther & Partners Ltd.

Figure B-3 Schematic of an electrical resistivity particle count sensor

Electrical resistivity sensors (see Figure B-3) require pumping the sample through a small orifice in a separation wall between two liquid chambers. One of the chambers is typically a tube immersed in a much larger container. Two electrodes, one inside the tube and another in the larger container, measure the liquid conductivity. Any time a particle passes through the orifice, the electrical flow through the solution reduces as the particle interferes with the electrical flow. Each occurrence of this interference is counted as a particle. The degree to which a particle reduces the flow of electricity is proportional to the particle size. This technique is appropriate for seawater or very brackish fresh water. Its application to fresh water requires increasing the water conductivity, for example, by dissolving salt in it.

Particle counting analysis can be applied to water samples using batch instruments, or to continuous sample streams using on-line instruments. Operators use two types of on-line instruments. True on-line systems (see Figures B-4 and B-5) apply particle counting to streams flowing from the main process, which are wasted after measurement. This technique does not allow the sample to be saved for any later use.

Batch/on-line systems (see Figure B-6) are similar, but particle counting is not a continuous process, although the sample stream flows all the time through the

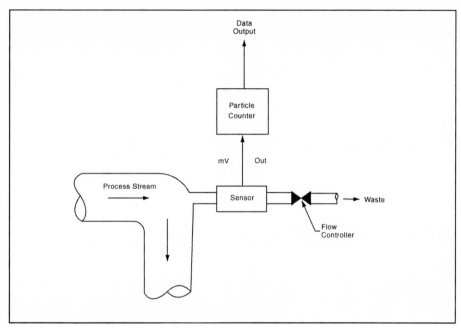

Source: *Reid Crowther & Partners Ltd.*

Figure B-4 Configuration for on-line particle counting

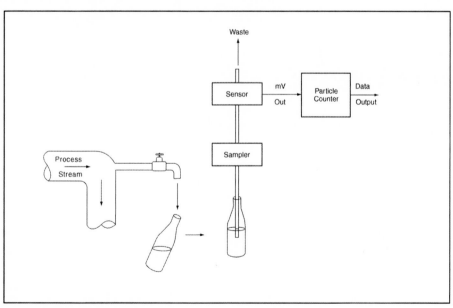

Source: *Reid Crowther & Partners Ltd.*

Figure B-5 Alternate configuration for on-line particle counting

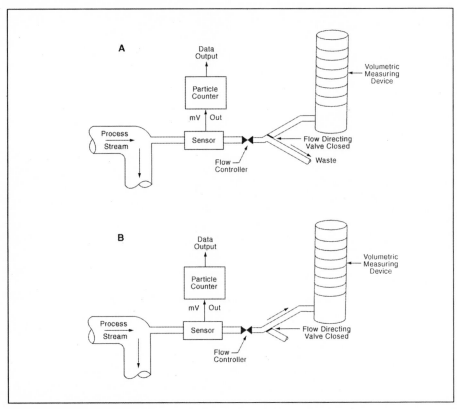

Source: Reid Crowther & Partners Ltd.

Figure B-6 Configuration for on-line particle counting: (A) Between counts and
(B) During counting

sensor. Instead, counting starts when the sample stream is diverted from waste and
into a sample container. Counting stops when the desired volume has been
examined, and the sampled stream is diverted back to waste. This method allows
better accuracy in the calculation of particle concentration and also permits saving
the sample.

Glossary

A

absorption fully drawing in (as a sponge absorbs water)

acid a compound that yields positive hydrogen ions in a water solution; a strong acid is one that yields a high percentage of hydrogen ions in a water solution

adsorption adhesion of the molecules of a gas, liquid, or dissolved substance to a surface. In water treatment, for example, manganese adsorption refers to manganese in solution (Mn^{2+}) adhering to the surface of manganese dioxide, $MnO_2(s)$, molecules.

ammonia a light, colorless gas, NH_3, with an irritating and pungent odor. It normally has a chemical demand for chlorine of almost 8 times its own mass (i.e., 8:1).

anion a negatively charged ion attracted to the positive pole in electrolysis

anthracite a type of coal with characteristics that often make it desirable for potable water filtration. Material mined in Pennsylvania is not the only anthracite coal suitable for use in potable water filtration. This handbook refers to *filter coal*, which may or may not be anthracite. Other commonly known coal types are bituminous coal and lignite coal. Bituminous coal from some deposits can be used for filtration

aquifer an underground layer of porous rock, gravel, or sand containing water

atom the smallest particle of an element that retains its characteristics

attrition the act or process of wearing away or grinding down by friction, or any gradual wearing or weakening, especially to the point of exhaustion

B

bench testing small-scale testing of chemical reactions or treatment processes on a bench, counter, or table

biofilm the term used in microbiology to describe bacteria slimes or plugs. It is also used to describe the bacteria covering the granular filter media in a biofilter for removing Fe and/or Mn.

C

carcinogen any substance that causes cancer

catalyst (catalytic agent) a substance that increases the speed of a chemical reaction without itself being permanently changed

cation (pronounced kat'i'en) a positively charged ion attracted to the negative pole in electrolysis

caustic soda a common name for sodium hydroxide, NaOH

Celsius scale the thermometer scale on which the freezing point is 0° and the boiling point is 100° at standard pressure

centimetre one one-hundredth of a metre

chemical change a change in which one or more new substances are formed

clean bed head loss (See also *head* and *head loss.*) head loss in a filter when the filter bed is clean. If the decision to backwash is based on head loss, then automatic equipment is triggered by the differential between the clean bed head loss and the dirty bed head loss.

clear well the place where water is stored before being pumped through the distribution system. In many plants, the clear well is a storage area under the facilities; water goes from there to enclosed storage reservoirs or water towers, then into the distribution systems. The clear well may be the only storage reservoir in the system.

coagulant a substance that triggers formation of a soft, semisolid mass in water, to which constituents to be removed are attracted and/or trapped by adhesion; often the constituents become heavy enough to settle out. Many such coagulants aid Fe and Mn removal, but the most common are alum and polymers of many types. Coagulants work by neutralizing or destabilizing the surface charge on the particulate to be removed, permitting flocculation to take place. (See also *flocculant.*)

cohabit (cohabiting) two or more sharing the same environment. For the purposes of this handbook, *cohabit* refers to microbiological organisms.

colloid a dispersion of particles larger than those in true solutions and smaller than those in true suspensions. In practical use, a colloid is a particle so fine it will not be filtered out by granular filter media (even by 0.30–0.35 mm manganese greensand). To filter colloids, a coagulant/flocculant first induces the particles to bunch up into units large enough to be filtered. The colloids most commonly removed from potable water carry negative surface charges.

complexing See *organic complexing.*

compound, chemical a substance composed of two or more elements chemically combined in definite proportions. For example, H_2O is a compound made up of two hydrogen and one oxygen atoms.

conditioning (See also *regeneration*.) usually, a reference to the act of applying a chemical in solution to a new, unused product, such as manganese greensand, to transform the material into a condition in which it can do its job. Before the $MnO_2(s)$ coating on manganese greensand can adsorb optimum amounts of Mn in solution, it must first be conditioned or regenerated with a strong oxidant, typically potassium permanganate $(KMnO_4)$ or chlorine (Cl_2).

cost–benefit analysis study to consider specifically both the cost and the benefit of doing something. For example, the cost to chlorinate water is $X per 1,000 m^3; the benefit is drinking water free of disease-causing organisms. Because the public demands safe drinking water, the benefit exceeds the cost.

D

decomposition a compound breaking down into simpler substances or into its elements. Also, rot or decay.

delta P pressure differential. Usually written using the symbol for delta in the Greek alphabet, Δ, followed by a capital P: ΔP

density weight per unit volume, as the grams per cubic centimetre of a solid, the milligrams per litre of a liquid, or the grams per litre of a gas

dewatering the removal of water from something. For example, in lime softening processes, large volumes of lime sludge accumulate. In order to get rid of the lime sludge, it is dewatered and then handled as a solid or semisolid by-product

disinfection making water safe for humans to drink, normally by adding chlorine, chlorine dioxide, or ozone, or by applying ultraviolet light rays

dosage the amount of chemical added to raw water during water treatment, usually expressed in milligrams per litre (mg/L). For example, 0.64 mg/L chlorine (Cl_2) helps to oxidize iron; other dosages might be stated as 1 mg/L fluoride, 5 mg/L sequestrant, etc.

E

eductor a device used to place filter media into a filter. The granular filter medium (e.g., coal or sand) is placed into a hopper, on the bottom tip of which, a venturi device transmits water under pressure (485–690 kPa, 70–100 psi) through a hose, creating sufficient suction to draw the filter medium out of the hopper into the hose, where it forms a slurry for delivery to the filter vessel.

effective size the sieve size such that 10 percent by weight of a granular media is smaller, often shown as e.s. or ES. For example, manganese greensand has an ES of 0.30–0.35 mm, or a mean ES of 0.325 mm. This statement means that 10 percent by weight is smaller than 0.325 mm, and 90 percent by weight is larger than 0.325 mm.

electrolysis decomposition of a substance, melted or in solution, by application of an electric current

electrolyte a substance that conducts an electric current when melted or in solution; also, a substance that ionizes in water

electrovalence the process in which metals give up electrons and nonmetals take on electrons when they combine to form a compound

element the simplest form of matter that can be obtained by any ordinary means, such as iron, oxygen, etc

exchange medium the common term for natural and synthetic zeolites. Although exchange media are used to remove a variety of elements and compounds from industrial process water, in potable-water treatment, the term describes the medium used to soften water using the cation exchange process. (See also *zeolite.*)

F

filter coal within this handbook, a comprehensive term for all coals used in potable-water filtration

filter startup the time when water starts flowing through a filter, including the first time ever, the first time following the installation of new filter media, the start of a new filter run following backwashing, the time at which a filter automatically returns to service, or any other time the filter begins service

fines dust and other small particles resulting from the manufacturing process for granular filter media. All filter media produced by crushing contains fines, which must be removed after installation in a filter vessel and before the filter is put into service. Particle breakdown (attrition) within the filter can also contribute to the level of fines.

finished water water ready to enter the distribution system

flocculants a group of chemicals that create flocs composed of themselves and aggregates of coagulated smaller particles that can be either settled out or removed by granular media filtration. Flocculation results from a large number of collisions between coagulated particles. (See also *coagulant.*)

fluidization bringing about the state in which a granular filter medium behaves as a fluid because a sufficient volume of backwash water flows through the filter bed. Generally, if a prodder (yardstick, copper pipe, rake, shovel handle, etc.) can be easily forced to the bottom of a filter bed during the backwash cycle, the bed is said to be fluidized. The main factor influencing the degree of fluidization is the volume of backwash flow

formula, chemical an expression containing the symbols of the atoms and subscripts (the numbers written below the baseline) showing how many of each kind of the atoms are present in the molecule; also the abbreviation for the molecule of an element or a compound. Examples include O_2 for oxygen gas or H_2O for the water molecule

G

granular filter media sand, crushed quartz, garnet sand, manganese greensand, and filter coal. Typical sizes range from 0.25 mm to 1.20 mm, though some sizes can be larger or smaller than this range. Water treatment uses other granular filter media, but this list cites those most often used.

H

hard water water containing mineral ions that react with ordinary soap to form insoluble substances. The most common hardness elements are calcium and magnesium (not to be confused with manganese). Stated simply, more softening allows water to more easily generate soap lather.

head the amount of energy possessed by a unit quantity of water at its given location (See the discussion of hydraulics in chapter 3.)

head loss a reduction in head. The source of head in a water treatment plant, i.e., the water level, is not very flexible, and therefore head loss is an important consideration. The major loss of head in a filtration system is caused by the filter media itself, since friction results from restrictions in flow. Head loss increases as these restrictions increase or the flow passages diminish in time. Head loss varies with the depth, shape and size, and cleanliness of the media. As the media becomes dirty due to accumulation of trapped particulate material (including oxidized iron and manganese), head loss grows (indicated by a rising reading on the pressure gauge on the raw water line feeding into the filter, due to an increase in pressure needed to force water through the filter). Thus, head loss is a means of determining when backwashing is required. Head loss over 2.5 m (8 ft) is generally accommodated in the design of a gravity filter.

hydrate a compound formed by a chemical combination of water and some other substance in a definite molecular ratio. Plaster of Paris, $CaSO_4 \cdot 2H_2O$, is a hydrate.

hydraulic carrying capacity at a given pressure, the amount of water a pipe can carry before friction results in unacceptably high pump head loss. This calculation is best left to engineers. However, hydraulic (water) carrying capacity can clearly decline if deposits of oxidized Fe and Mn, bacterial growths, crud, silt, etc., build up in a potable-water distribution system. Such buildup reduces the inside diameter of the pipe and raises the electrical costs of driving pumps. The cheapest distribution line to operate is a clean one

hydrologic cycle the endless movement of waters from the sea to the atmosphere (by evaporation), from the atmosphere to the land (by rain, snow, sleet, and hail), and from the land back to the sea via streams, rivers, and lakes.

hypochlorite any salt of hypochlorous acid (HOCl)

I

in-situ in water treatment, the location of a process where it is specifically needed. For example, coating filter medium particles with $MnO_2(s)$ in the filter during operation, as compared to artificially coating particles in a factory somewhere away from the treatment plant. Another example is described in chapter 7, which discusses in-situ removal of Fe and Mn targeted at the raw water before it is drawn from the ground source (well); in other words, treatment *in-the-situation* where the Fe and Mn are first encountered.

inversions See *water inversions.*

ion an electrically charged atom or radical

ionization separation (dissociation) of a molecule in a water solution into electrically charged atoms or radicals called *ions*

ion exchange a chemical process for reversibly transferring ions between an insoluble solid, such as a resin, and a fluid mixture, usually a water solution. The exchange in water of the ions of one element for the ions of another. For example, sodium or potassium cations attached to a zeolite (as in a water softener) exchange themselves for cations of calcium and magnesium, softening the water passing through the zeolite bed. (See also *hard water; zeolite process.*)

M

$MnO_x(s)$ manganese oxides, including manganese dioxide—$MnO_2(s)$. Throughout this handbook, the formula for manganese dioxide is shown as $MnO_2(s)$ in order to be technically and chemically correct, since subtle but important differences distinguish MnO_2 from $MnO_2(s)$. However, the term *manganese dioxide* is considered acceptable common usage.

molecular weight the sum of the atomic weights of the atoms in a molecule. The molecular weight of a molecule compared to the carbon atom is 12, equivalent to the total numbers of protons and neutrons in the nuclei of the atoms forming the molecule. Science has determined the practical effect of molecular weight on organic complexing by certain carbon compounds, which can have an impact on designing processes for Fe and Mn removal.

O

$O_2(aq)$ oxygen in an aqueous solution, i.e., oxygen in water. Most commonly, oxygen is forced into water by simple aeration.

organic chemistry the chemistry of carbon compounds. The terms *organic complexing* and *organic binding* refer to the process through which such elements as Fe and Mn become part of a carbon compound. Chemists have identified literally thousands of carbon compounds.

organic complexing, organically bound, organic binding, carbon complexed in Fe and Mn removal, associations between the metals and organic carbon compounds

organic compound a compound in which carbon is the chief constituent. Originally, this term meant any compound obtained from living organisms.

oxidation in a limited meaning, a chemical combination with oxygen. In a broader sense, this term means the loss of one or more electrons (negative valence) from an element or a radical or an increase in positive valence.

oxidation–reduction a chemical reaction in which one reactant is reduced (gains one or more electrons) and another is oxidized (loses one or more electrons). The process based on this reaction is referred to commonly as *redox*. (See *redox*.)

P

polymer a general term for chemicals composed of long chains of molecules of known electrical charge and electrical strength. These compounds aid water treatment by agglomerating (clumping together in bunches) very small particles so that they can settle out of water and/or become trapped in filters.

potable water water fit for humans to drink

pyrolusite a mined solid particle of manganese dioxide that does the same job as the $MnO_2(s)$ coating on manganese greensand. It is much harder than the glauconite sand base of manganese greensand, so it will not break down under filter head differential pressures as low as 55 kPa (8 psi). A typical sand-pyro matrix (mixture) used in filtering contains 10–50 percent pyrolusite. It has been the $MnO_2(s)$ filter medium of choice for Fe and Mn removal in Great Britain for decades

R

radical a group of two or more elements that act as one in any chemical reaction. For example, in Na_2SO_4 (sodium sulfate), $(SO_4)^{2-}$ is a radical.

raw water water as it comes from the source (well, lake, reservoir, river), or untreated water

redox in a groundwater system, the microbial biomass (a growth of biological organisms) often focuses at the oxidation–reduction interface at Eh values from +150 to –50 mV. Since an oxidative (aerobic) state (with positive Eh values) will support aerobic microbial activities, while a reductive state (with negative Eh values) will encourage anaerobic activities, the range Eh +150 to –50 mV is the scientific description for the change from aerobic to anaerobic states. As Figure 7-2 shows, biological oxidation of iron can happen before physical–chemical oxidation. Figure 4-2 refers to mV, while Figure 7-2 cites V; the first unit is millivolts, or 1/1,000th of a volt. (See also *oxidation–reduction*.)

regeneration to produce (a compound, product, etc.) again chemically, as from a derivative or by modification to a physically changed, but not chemically changed, form. In water treatment, regeneration is the process by which chemically coated filter media such as manganese greensand [which is coated with $MnO_2(s)$], or water softener zeolite resin (a cation exchange medium) is returned or restored to its top-producing condition. To regenerate manganese greensand or pyrolusite, the regenerant used is either chlorine (Cl_2) or potassium permanganate ($KMnO_4$). Water softener zeolites are typically regenerated with either a water solution of sodium chloride or potassium chloride (both softener salts).

residual the amount remaining after a process has been completed. For example, a free chlorine residual of 0.5 mg/L is left over after all the other chlorine demands in a particular water have been satisfied. A Mn residual is the amount remaining in the water after the process has removed all it can just prior to a residual test.

S

sequestrant any agent or chemical capable of separating an element out of a compound. For example, polyphosphates can separate Mn from a compound of which it is part without changing the chemical description of that compound or the Mn sequestered. The "separating" is sequestration.

sessile attached directly to the main stem (See chapter 4 on microbiology.)

solute material dissolved in a liquid

sorption both or either of absorption and adsorption

specific gravity the density of a substance compared to the density of some other substance as a standard. Water is the standard for solids and liquids, while air is the standard for gases. For filter media, specific gravity (often shown as s.g. or SG) is the ratio of the mass of a volume of media to the mass of an equal volume of water under specified temperature conditions.

(ASTM Standard Test C128-84 specifies this temperature as 23°C.) The SGs of many filter sands average 2.60–2.65

static mixer typically, a pipe with vanes or baffles inside to generate turbulence in water flowing through it to promote total mixing of a chemical with a raw-water flow. A static mixer is often the same size as the raw water line. The word *static* means it has no moving parts.

stratification in this handbook, one of two occurrences: (1) Filter media remaining in a layer by itself or returning to such a layer following intermixing with other filter media by air scouring and/or water back-washing. For example, coal topping manganese greensand should remain in a layer above manganese greensand with little intermixing where the two media meet (also known as *interfacial mixing*). (2) A single filter medium separating out in layers based on particle sizes and densities (specific gravities). For example, as new greensand is thoroughly back-washed for the first time, comparatively large, heavy particles move to the bottom of the greensand layer, while the undesirable fine particles move to the top of the layer, from which they can be manually removed.

T

temporary hardness equivalent to alkalinity, if alkalinity is less than total hardness

total dissolved solids (TDS) the mass of ions plus silica

transpiration the method by which plants take moisture from the soil and transmit it to the atmosphere

turbidity solid particles in a given volume of water. Most turbidity meters interpret particle densities as NTUs (nephelometric turbidity units). Water with a high turbidity value has a cloudy or unclear appearance; turbidity of drinking water should not exceed 1.0 NTU.

U

uniformity coefficient a dimensionless factor to describe the uniformity of particle size in a granular filter medium. It is defined as the sieve size that passes 60 percent of the media grains by weight divided by the sieve size that passes 10 percent of the media grains by weight. Also shown as u.c., UC, and sometimes U_c. A UC of 1 would mean that all particles are exactly the same size (a situation not possible through commercial processes). Uniform filter sand has a UC of 1.3 or less, while a UC of 1.7 is considered lacking in uniformity.

V

valence the number of electrons that an atom or radical can lose, gain, or share with other atoms or radicals. *Valence* means essentially combining power; it is the relative worth of an atom of an element in combining with the atoms of other elements to form compounds. The valence number of an element is the number of its electrons associated with formation of a particular compound. For example, the valence of Mn^{+6} is 6.

viscosity a term from physics that describes a liquid's thickness. Formally, it is the internal friction of a fluid, caused by molecular attraction, that makes it resist flowing. The viscosity of water is influenced by temperature. Water is at its maximum viscosity (its thickest or maximum density) at $4.1^{\circ}C$ ($+39.4^{\circ}F$). Temperature affects the viscosity of water to such an extent that filter backwash rates routinely take into account the water temperature. See Figure 6-15.

W

water inversions In much of North America, temperature changes from one season to another result in significant water temperature changes in impoundments (reservoirs and lakes). In a reservoir, for example, the fall temperature of the surface water declines as the weather gets colder, while the water at the bottom of the reservoir remains warmed by the earth. When the surface water goes down to $4.1^{\circ}C$ ($39.4^{\circ}F$), that layer of water reaches its maximum density and slowly begins to sink, displacing the water at the bottom, which is forced to the top of the reservoir. This happens again in the spring after the ice melts. If the water at the bottom of the reservoir is anaerobic (i.e., lacking oxygen), it will have Fe and Mn in solution, which it will carry to the surface. There, mixing with oxygen from the atmosphere by wind and wave action slowly oxidizes the Fe and Mn forming solid precipitates that fall to the bottom of the reservoir again. This oversimplified explanation omits other chemical and biological occurrences. A plant operator must notice that Fe and Mn can change locations within a water body as the result of inversions, which affects the treatment process. The water rising from the bottom of a reservoir brings odors with it, usually described as earthy, septic, musty, or smelling like rotten eggs.

Z

zeolite any of a large group of natural hydrous aluminum silicates of sodium, calcium, potassium, or barium, chiefly found in cavities of igneous rocks and characterized by a ready loss or gain of water by hydration; many are capable of ion exchange with solutions. Most water softening resins are synthetic products, often loosely referred to as *zeolites*, though they are not.

zeolite process a water-softening process that removes ions causing hard water by passing it through a filter containing zeolite (a complex silicate)

References

American Public Health Association, American Water Works Association, and Water Environment Foundation. 1998. *Standard Methods for the Examination of Water and Wastewater*, 20th ed. Washington, D.C.: APHA.

Canadian Centre for Occupational Health and Safety. 1985. *Chemical Hazard Summary*. Hamilton, Ontario: CCOHS.

Cullimore, D. Roy. 1993. *Practical Manual of Groundwater Microbiology*. Chelsea, Mich.: Lewis Publishers.

Hach *Water Analysis Handbook*, 2nd ed. 1992. Loveland, Colo.: Hach.

Inversand Company n.d. *Manganese Greensand Cr & IR*. Brochure.

Hargesheimer, Erika, et al. 1992. *Evaluation of Particle Counting as a Measure of Treatment Plant Performance*. Denver, Colo.: AWWA Research Foundation.

Knocke, W.R., S.C. Occiano, and R. Hungate. 1991. Removal of Soluble Manganese by Oxide-Coated Filter Media: Sorption Rate and Removal Mechanism Issues. *Jour. AWWA* 83(8): 64–69.

Knocke, W.R., J.E. VanBenschoten, M. Kearney, A. Soborski, and D.A. Reckhow. 1990. *Alternative Oxidants for the Removal of Soluble Iron and Manganese*. Denver, Colo.: AWWA Research Foundation and AWWA.

Mouchet, Pierre. 1992. From Conventional to Biological Removal of Iron and Manganese in France. *Jour. AWWA* 84(4): 158–167.

Prescott, L.M., J.P. Harley, and D.A. Klein. 1990. *Microbiology*. Dubuque, Ia.: Wm. C. Brown.

Voorinen, A., et al. 1988. Chemical, Mineralogical, and Microbiological Factors Affecting the Precipitation of Fe and Mn from Groundwater. *Water Sci & Tech* 20(3): 249.

Additional Sources

Amirtharajah, A., N. McNelly, G. Page, and J. McLeod. 1991. *Optimum Backwash of Dual Media Filters and GAC Filter-Adsorbers With Air Scour*. Denver, Colo.: AWWA Research Foundation and AWWA.

Coffey, B.M., W.R. Knocke. 1990. Removal of Soluble Iron and Manganese from Groundwater by Chemical Oxidation and Oxide-Coated Mixed-Media Filtration. In *Proc. AWWA Annual Conference*. Denver, Colo.: AWWA.

Hill, M.T., M.C. Rand, and J. O'Brien. n.d. *Manual of Instruction for Water Treatment Plant Operators*. Albany, N.Y.: New York State Department of Health, Office of Environmental Health Manpower, Divison of Pure Waters, Office of Public Health Education.

Knocke, W.R., L. Conley, and J. VanBenschoten. 1990. Impact of Dissolved Organic Carbon on the Removal of Iron During Surface Water Treatment. In *Proc. AWWA Annual Conference*. Denver, Colo.: AWWA.

Knocke, W.R., H.L. Shorney, and J. Bellamy. 1993. *Impacts of Dissolved Organic Carbon on Iron Removal*. Denver, Colo.: AWWA Research Foundation and AWWA.

Knocke, W.R., H.L. Shorney, and J.D. Bellamy. 1992. Reactions Between Soluble Iron, Dissolved Organic Carbon, and Alternative Oxidants During Conventional Water Treatment. In *Proc. AWWA Annual Conference*. Denver, Colo.: AWWA.

Lewis, C.M., E.E. Hargesheimer, and C.M. Yentsch. 1992. Selecting Particle Counters for Process Monitoring. *Jour. AWWA* 84(12): 46–53.

White, G.C. 1986. *Handbook of Chlorination*, 2nd ed. New York: Van Nostrand Reinhold.

Index

Note: f. indicates figure; t. indicates table.